George Henry Lewes

Sea-side Studies at Ilfracombe, Tenby, the Scilly Isles, and Jersey

Second Edition

George Henry Lewes

Sea-side Studies at Ilfracombe, Tenby, the Scilly Isles, and Jersey
Second Edition

ISBN/EAN: 9783744730495

Printed in Europe, USA, Canada, Australia, Japan

Cover: Foto ©ninafisch / pixelio.de

More available books at **www.hansebooks.com**

AT

ILFRACOMBE, TENBY, THE SCILLY ISLES, AND JERSEY

BY

GEORGE HENRY LEWES

AUTHOR
PHYSIOLOGY OF COMMON LIFE, LIFE OF GOETHE

With Illustrations

"C'est surtout en histoire naturelle qu'on est toujours mécontent de ce qu'on fait, parceque la Nature nous montre à chaque pas qu'elle est inépuisable."
— CUVIER

SECOND EDITION

WILLIAM BLACKWOOD AND SONS
EDINBURGH AND LONDON
MDCCCLX

TO

OUR GREAT ANATOMIST

RICHARD OWEN

This Work

IS INSCRIBED

PREFACE.

THE substance of these pages originally appeared in *Blackwood's Magazine* during the years 1856 and 1857. They are now republished with considerable alterations and additions, which, it is hoped, will render them more useful to the amateur and student of Natural History.

I have endeavoured to furnish the visitor to the sea-side with plain directions, by means of which he may study and enjoy the marvels of ocean-life; and to present such descriptions of the animals, and the wonders of their organisation, as may interest the reader by his own fireside. With regard to the former, having had to ascertain almost everything for myself, I have tried to make my experience available for others; and the remembrance of early difficulties has suggested the statement of many details which to the well-informed may appear trivial, but for which I should myself have been very grateful. The papers in *Blackwood* were chiefly written at the coast, where the command of books was necessarily small, and the literature of each subject could only be given sparingly; this deficiency has now been supplied, as far as my resources extended; but no gathering of second-hand references has been allowed to make the labour seem greater than it really was.

A Glossary of technical terms has been drawn up for the general reader; and this, together with the Illustrations, will, I hope, render the descriptions perfectly intelligible, even to readers unfamiliar with marine animals.

There are two aspects in which Natural History may be regarded—as an amusement, and as a science; the one being simply delight in natural objects, the other a philosophic inquiry into the complex facts of Life. As this latter was the motive which prompted my visits to the coast, it has naturally assumed a prominent place in these pages; and although, throughout, a style of popular exposition is adopted, which aims at being intelligible to all cultivated readers, I have also had a special audience in view, to whom must be submitted the appreciation of the new facts and new physiological interpretations herein advanced. Every competent person will see that these novelties are the result of hard work and continuous application; and I have been careful to indicate the amount and kind of evidence—Observation, Dissection, or Experiment—on which they rest; so that the source of error, wherever there is error, may be detected.

The most startling of the new views, that, namely, on the identity of Growth and Generation, has recently received striking confirmation in the admirable researches of Professor Huxley on the Aphides—researches which render the facts I have observed in the development of Polypes less paradoxical than they originally appeared.

January 1858.

CONTENTS.

PART I.—ILFRACOMBE AND TENBY.

CHAPTER I.

Page

The start—Ilfracombe—How to hunt for marine animals—Necessary equipment—A day's hunt—Sea-hares—Terebellæ—Aspects of the sea—Return home—Glass jars and glass tanks—Identifying animals—Books necessary for the student — Delights of the microscope — Popular errors respecting the microscope—Experiment, ... 2-41

CHAPTER II.

Superior charms of marine-hunting—Morning labours—The hermit-crabs—Corsican brothers—Food of the crab—A comical fish—Zoological paradoxes—Study of nature—Vitality of separated parts—Definition of life—Organic law of division of labour—The terebella: its spontaneous fission (Bonnet's experiments), its two circulating fluids, and double respiration—Structure and function of its tentacles—Erroneous use of the terms muscle and nerve—Always carry a basket—The lanes of Ilfracombe, 42-81

CHAPTER III.

Live animals sent by post—The Morte Stone—"Wanted, a tin canister!"—The boring Molluscs, Pholas, Teredo, and Saxicava—How is the boring effected?—Application of embryology to practical purposes—Respiration of the Pholas—Dredging—The sea a passion—The dead Cephalopod and its colour-specs—Need of a doctrine—Mormon preacher—Logic of Zoologists—Cleopatra's pearl—Produce of a gale—The Eolis: its structure; how does it breathe? reasons for removing the Eolis from the nudibranchiates—Water mingling with the blood, ... 82-118

PART II.—SEA-ANEMONES.

CHAPTER I.

Sudden enthusiasm for the Anemones—Imperfect state of our knowledge—Literature of the subject—No distinction between animal and plant: sensitiveness, locomotion, and capture of food—Voracity of the Anemone—Its paralysing power denied—Habits and instincts—Is it viviparous only?—Variations of colour, ... 121-151

CHAPTER II.

Desires for abundance—The thread-capsules: are they nettling organs?—Futility of observation uncontrolled by experiment—Structure of an Anemone—General law of development—Development of the human hand—Reproduction of Anemones—Their ovaries and spermatozoa—Are Anemones of separate sexes, ... 152-184

PART III.—THE SCILLY ISLES.

CHAPTER I.

The lion that has eaten a man, and the zoologist who has been at the coast—Troublesome desires—Choice of the Scilly Isles—Penzance lodgings—The sail to Scilly: pursuit of knowledge under difficulties—First sight of the islands—Their area and population—Their picturesqueness—The changeableness of rocks—Antiquities of Scilly—The inhabitants—Primitive state of the commerce—Dinner difficulties—How the ten thousand saluted the sea—Love of the English for the sea—Homer—Our first day on the rocks—The Nymphon Gracile—The Comatula—On observation and experiment in biology—Do the Anemones digest?—Meaning of digestion—Assimilation and digestion—The Actinophrys—Food and blood—Experiments on the Anemones—Food and knowledge, . . 187-232

CHAPTER II.

Druidical remains—A wreck off Scilly—Geology and Zoology of the isles—Effect of light on plants and animals: history of its discovery—The Pipe-fish and its incubation—Fish paradoxes—An aquarium—Suicide of the Starfish—The pleurobranchus—Development of Eolis, Doris, and Actæon—Shell and no shell—The Pedicellina: is it viviparous?—The Sagitta: a puzzle to zoologists—Where there is no respiration there will be no circulation—The chylaqueous fluid of Anemones proved not to exist—Earliest stage of a nutritive fluid—Function of the "convoluted bands"—Delights of literature, ... 233-279

CONTENTS.

PART IV.—JERSEY.

CHAPTER I.

Page

Departure from Scilly—Jersey—Schoolboy recollections—Choice of a spot—Prejudices against dissection—Zoology of Jersey—Trawling—Difficulties of identification—A strange Ascidian—The Tadpole embryo of a Mollusc—A paradox—Development of Medusæ from Polypes—All animals do not come from eggs—Reproduction of the Hydra—History of Parthenogenesis—A discovery—Animals produced from germ-cells only—Steenstrup's theory—Owen's theory—A new theory—The three forms of reproduction: fissiparous, gemmiparous, and oviparous—What is parthenogenesis? 283-340

CHAPTER II.

Summer delights—Medusæ-hunting—Noctiluca and the phosphorescence of the sea—The Cydippe—Vivisections—Do the lower animals feel pain?—Chance-weed—A new Polype—A new Polyzoon—Vitality of Molluscs—Vision of the Molluscs—Are images formed on the retina?—Description of the retina in vertebrates and invertebrates—New theory of vision—Tactile sensations and nerve filaments—Can the Molluscs hear?—The senses of animals not superior to those of man—The ocean-currents caused by Molluscs, ... 341-379

CHAPTER III.

How to catch Razor-fish—The Corkscrew Coralline—Danger of *a priori* views in zoology—Examination of the nervous system of Molluscs——Doubts respecting current doctrines of nerve-physiology—Absence of nerve-fibres—Origin of sensibility—Sensibility in the absence of nerves—On the relation between organ and function—Conclusion, . . 380-412

GLOSSARY, . . 413

INDEX, ... 417

DESCRIPTION OF THE PLATES, . 426

PART I.

ILFRACOMBE AND TENBY.

SEA-SIDE STUDIES.

CHAPTER I.

THE START—ILFRACOMBE—HOW TO HUNT FOR MARINE ANIMALS—
NECESSARY EQUIPMENT—A DAY'S HUNT—SEA-HARES—TEREBELLÆ
—ASPECTS OF THE SEA—RETURN HOME—GLASS JARS AND GLASS
TANKS—IDENTIFYING ANIMALS—BOOKS NECESSARY FOR THE
STUDENT—DELIGHTS OF THE MICROSCOPE—POPULAR ERRORS
RESPECTING THE MICROSCOPE—EXPERIMENT.

A FEW warm sunny days in April 1856, which flew over our heads like swallows twittering of the coming summer, stirred in my bosom irresistible longings to quit the moil and turmoil of metropolitan crowds for the bright and breezy coast. As I hurried through the noisy London streets, or sauntered through the comparatively quiet lanes of Richmond and Twickenham, and looked at the summerlike sky above, I began to understand the migration of birds, and felt something of what they must feel, when certain dim imperious influences urge them to quit their homes, and traverse many a weary league of foam, in search of the resting-place awaiting them somewhere with a warmer smile. I grew impatient to take wing; and asked myself constantly—

> "Why, though ill at ease
> Within this region I subsist,
> Whose spirits fail within the mist
> And languish for the purple sea?"

Indeed, I languished for the sea with a sort of zoological calenture. Ill health was but a minor pretext; the major pretext was an ever-accelerating desire to become more intimately acquainted with the organisation of marine animals, as an aid in those researches in General Physiology which, commenced some twenty years ago, had for the last six years become a passion.

The sea was an old playfellow of mine, and from boyhood our friendship had been continually heightening. It is almost needless to add that, like most boys who live on the coast, I had early visions of being a sailor; to be a tar seemed the culmination of all earthly ambition. Indeed, as a preparatory training to that high aim, at nine years old I declined to wear gloves, and secretly chewed bits of cigar, in guise of a quid, as an evidence of nautical qualification. The tobacco was horrible; but manliness at nine years old disdains nausea, and I "stood" to my quid with the unflinching resolution to be afterwards shown in standing to my guns. To rehearse the intensely nautical flavour of my diction at that period, and my fatal accuracy in all technical terms, is beyond my power. But whatever else I learned of the sea, I learned little of its inhabitants. Starfishes and jellyfish could not fail of being observed. Crabs are hunted and tortured by all amiable boys; and the bone of the cuttle-fish duly prized. But in regard to its marvels and its inhabitants, the sea was a new acquaintance to make. Mr Gosse and others had made marine zoology fashionable. The Aquarium at Regent's Park had given glimpses of the wonders and the beauties which abound on the shore; and as a student of Comparative Anatomy, I had been led to read extensively respecting the structure and functions of marine animals; but of direct knowledge I had next to nothing. I had learned something of ponds and their inhabitants, and was now desirous of ransacking the sea.

Ilfracombe was the spot fixed on: a more charming spot England could hardly furnish. At first I knew not how much of the delight with which its beauty thrilled me might be owing to the mere effect of comparison and novelty. After the metropolis, any broadening blue of sea, any bold headland or straggling reef, seems supremely beautiful—and novelty is, in itself, an integer in all travelling enjoyment. But familiarity only served to deepen my sense of the beauty of Ilfracombe; the very last look was taken with a reluctance springing from unsatiated desire; and on reaching Tenby, also a charming spot, the overpowering sense of disappointment assured me that Ilfracombe *was* the enchantress she had seemed. I will not describe Ilfracombe, and for two reasons: First, it would occupy all the chapter; and, secondly, which perhaps is as good a reason as the other, I have no descriptive talent. Had

I the talent, the picture would be tempting, for the charms of the place are manifold. The country all round is billowy with hills, which rarely seem to descend into valleys. The paradox may move your scepticism; you may bring excellent reasons, physical, geological, and geographical, to prove that wherever there are hills there must be valleys. Nevertheless, the abstract force of *what must be* vanishes before the concrete force of *what is;* and at Ilfracombe you will find hills abounding, hills rising upon hills, but not always making valleys. What the French picturesquely call the *mouvement du terrain,* which suggests hills in motion like the waves, is here seen on every side; and these waving slopes are in spring-time pale with primroses, or flaming with furze. If you get sight of a bit of earth to vary the verdure, it is of that rich red-brown marl which warms the whole landscape. If you climb one of those hills, the chances are that you come upon a rugged precipice sheer over the sea; unless a green slope leads gently down to it. These breezy hills, and the soft secluded valleys (there *are* valleys), and the matchless lanes which intersect the land with beauty, afford endless walks of varied delight. The lanes of Devonshire are celebrated; but what Shakespeare's works are to the criticisms which celebrate them, these lanes are to their reputation. Were I to enter one of them, and begin describing it, we should never get down to the shore, whither I see your impatient footsteps tend. To the shore, then! and as we pass, we can take a glimpse of the town.

Handsome the town of Ilfracombe is not; nor, although picturesquely placed, has it a very picturesque appearance, except under certain lights, and from certain points. The colour of the houses is pale dingy grey; the lines are all rectangular and mean. Overtopping the whole town in ugliness and pretension, no less than in altitude, are two terraces, which make two factory-like lines of building on the slope of the green hill. You see at a glance that the flounces and shaved poodles live there.

Yet, as I said, there were lights under which the town looked well; but what will not light transform into beauty? One evening, after a shower, I was called away from the Microscope to look at the town under the light of the setting sun, some peculiar arrangement of the clouds, with a vivid rainbow, having thrown a delicious evening tinge over the houses piled on the sides of the hill, and merged the ugliness of their forms in exquisite floods of colour. In this light Ilfracombe looked handsome. It looked resplendent, like a stupid man in the splendour of a noble deed.

If unblessed with the fatal (but agreeable) gift of beauty, the little town of Ilfracombe, as a compensation, is uncursed with the appearances of pretension. Except on those two unfortunate terraces, it gives itself no airs of fashion, no demure hypocrisies of respectability. It has no magnificent hotels; it has no popular preacher. It makes nobody miserable. *Simplex munditiis;* a plain face, but clean and honest, sirs! I was continually reminded of some small German town, and the simple honesty and obligingness of the people helped the resemblance. As you enter from the Braunton road, there is a white-washed inn, now untenanted, of the most primitive structure, and bearing the words

BRAUNTON INN

painted in tall brown letters all along the frontage, which I never passed without some vague reminiscence of Germany rising up, so exactly does this turn of the road repeat many turns of road I have come upon in my wanderings. An avenue of mountain-ash, with their bright red clusters brilliant against the hot blue sky, or rows of plum-trees with their purple fruit, pleasing the eye and refreshing the palate during the dusty walk, would have made the illusion complete.

Let us pass this inn, and turn up the steep hill, on the summit of which stands the handsome church; we then descend the slope which leads to the Baths. On the other side of the hedge, upon our left, rise the soft uplands; and a little behind them the majestic Seven Tors, which with their shaggy heads towards the sea, and their soft-swelling slopes of green towards the land, remind us of some mighty animal which has reared itself on its fore-paws to gaze at the yet mightier ocean. From these uplands you perpetually hear the cry, day and night, of the landrail — just like the creaking of a wicker-basket — so that you begin to wonder when that unmusical bird takes its repose.

On your right hand, the clear Wilder-stream babbles incessantly to the wild-flowers nodding over its ripples. Accompanied by this music we reach the Baths, and come upon a tunnel, dark, indeed, but with a gleam of light at the end. We enter: how cool, not to say cold! The eye is getting familiar with the darkness when we emerge, and what a *thrill* runs along all the sentient paths to our souls as the blue of the sea bursts upon us! We lean upon a parapet of rock to watch the waves running up the rugged face of the cliffs, and falling back in spray. An inarticulate gasp does duty for the highest eloquence. It is enough to drink in with our eyes the scene before us; anything more than an incoherent exclamation would be out of place. Another tunnel invites us; through it we pass, and come upon a wooden bridge overarching an ugly-looking spot bearing the name of Tracy's Cave, which has of course its devout legend to tell, if you are willing to listen. Let the legend be what it may, the place is grim, and at first we tread cautiously as we pass over the bridge of logs; but soon familiarity reconciles us to this—as it does to small-pox or the income-tax. Before reaching this spot we have come upon another opening, leaned upon another parapet, and had another gaze at the sunset gleaming over the sea. We now step on the wild and rugged shore.

And what a shore! Precipitous walls and battlements of rock rise on each side, making a bay; before us, sharply-cut fragments of dark rock start out of the water for some distance. Every yard of ground here is a picture. The whole coast-line is twisted and waved about into a series of bays and creeks, each having a character of its own; and whether we stand on the Tors, and look along the coast—or on the shore, and look up at the rocks, there is always some new aspect, something charming for the eye to rest upon.

> " An iron coast and angry waves,
>   You seem'd to hear them climb and fall,
> And roar, rock-thwarted, under bellowing caves,
>   Beneath the windy wall." *

The rock is *grauwacke* or clay-slate, with occasional streaks of quartz; and the stratification is very various. Look at that reef, round and along which the stealthy tide is crawling; see how the back of it is ridged with sharp sudden lines cutting against the sky; or look at that sombre precipice over which the gull is floating broad-winged, uttering its piteous cry, or startling you with its strange mocking laugh. A little farther and the eye rests on a purple-tinted wall of rock, from the sides of which jut ledges covered with vegetation. The soil here is so generous, that nature seems to be bursting into life through every crevice, and on every inch.

There is, however, one serious drawback at Ilfracombe—the complete absence of sands whereon to loll or stroll, or, in the quiet hours of moonlight, to wander nourishing one's middle age sublime " with the fairy tales of science and the long results of time." However, sands are poor hunting-grounds; let us take con-

* Tennyson.

solation in that, and enjoy the positive excellences of this place.

The evening of my arrival was spent in reconnoitring the coast and its promises. What a flutter agitated me as I bent over the many rock-pools, clear as crystal, and sometimes enclosing perfect landscapes in miniature. It seemed as if I should have nothing to do but stoop and fill my jars with treasures; for I had read in numerous books descriptions from which the inference was, that nothing could be easier than collecting "marine store." "You stroll along the beach and pick up so-and-so," is the pleasant phrase of these writers, wishing, we must suppose, to make science appear easy. Now the truth should be told. It was quickly forced on my conviction that, although after a gale you may go down to the shore and find many things, mostly dead, which you will carry home with interest—for "'tis an ill wind that blows nobody molluscs"—yet hunting among the rocks is not easy, nor always safe, nor certain to be successful. You must make up your mind to lacerated hands, even if you escape bruises, to utter soakings, to unusual gymnastics in wriggling yourself into impossible places. You can only do this successfully at certain tides. And, after all, you may return empty-handed, unless you are very modest in your desires. I did, indeed, behold a stout gentleman, who had been reading Mr Gosse, severely deluding himself into the idea that he was "collecting," because he was gasping among boulders with a pickle-bottle in one hand and a walking-stick in the other: but I am not firmly persuaded that he carried home much worth his trouble.

Let me mention the proper equipment for a day's hunting, and you will see that the pickle-jar and walking-stick theory is primitive, and somewhat inefficient. It is necessary to take with you a geologist's hammer (let it be of reasonable size), and a *cold* chisel; to these add an oyster-knife, a paper-knife, a landing-net, and, if your intentions are serious, a small crowbar. You will want a basket, which must be tolerably large, and flat-bottomed. Having made that small investment, you turn into the chemist's and buy up all the wide-mouthed phials he will sell — those used for quinine are the best. The short squat bottles, with wooden caps, now sold for tooth-powder, are very convenient. We lay hands on half-a-dozen of these, and having laid in three or four earthenware jars (not to be too abstract in our diction, let us frankly say jam-pots), we return home to construct our collecting-basket, which is done in this primitive fashion: A loop of string serves to keep a large jam-pot at one end of the basket, at the other end another loop sustains a large phial; at one side a loop is made for hammer and chisel; opposite are two more phials. Mr Gosse, in his *Aquarium*, I think, describes the basket he uses; but as you must order this to be made for you, my plan is sufficiently serviceable, and costs no trouble. The basket ready, we are now equipped. No; there is still one little implement. A piece of brass wire, the end twisted into a ring of two or three inches diameter, to which is fastened a canvass bag, makes a convenient little net to be used in pools too small to admit the landing-net.

The brief note in my journal which records the results of my first visit always amuses me when it catches my eye: "On the rocks. Found some Actiniæ and Serpulæ." The idea of *finding* Serpulæ will make the amateur smile as he remembers how difficult it is to avoid these swarming Annelids, whose shells, sharp as lancets, cut the hands in fifty different places before many stones are turned; but to my inexperienced eye there were only the empty shells of Serpulæ to be found, until I came upon some in the water with their little fans expanded, and these were pounced on with great eagerness. The Actiniæ spoken of were the common Smooth Anemone — not even the Strawberry variety—(if you will face a long name, it is *Mesembryanthemum*)—and these which I bagged with great glee, I soon learned to pass by with no more regard than if they had been sea-weed. So much of our enjoyment depends on the difficulty of obtaining it, that these Actiniæ, which I still hold to be exquisitely beautiful, and far more intrinsically beautiful than very many of the rare species, to obtain which one nearly dislocates one's limbs, wriggling through crevices, or runs a risk of "catching one's death" by standing in a pool dripped on from a thousand orifices above—these Actiniæ, I say, are left untouched because they are abundant, and do not demand the chisel. Perverse, ungrateful human nature! What should we not think of daylight, or of woman's patient love, if it were not given with such generous abundance? Ask the prisoner, or the man who has scarcely known the mother's ceaseless tenderness, the wife's surpassing love. The coquette knows this by instinct, and she draws adventurous seekers after her. What a coquette is the Daisy (*Actinia bellis*), who displays her cinq-spotted bosom, beautiful as Imogen's, in the crystal pool. You are on your knees at once; but no sooner is your hand stretched towards her, than at the first touch she disappears in a hole. Nothing but chiselling out the piece of rock will secure her; your labour is the price paid for the capture, and the captive is prized accordingly; if as much labour had been given to the Smooth Anemone, she would have seemed as lovely in your eyes.

There is something sad in the fugitive keenness of pleasure. I shall never feel again the delight of getting my first Actinia. No rare species can give that peculiar thrill. There is a bloom on the cheek which the first kiss carries away, and which never again meets the same lips. No partridge is worth the first which falls by your gun; no second salmon is ever landed with the same pride as the first. Even printer's ink has a perfume when your first "proofs" arrive. Who will revive within me *that* flutter which deprived me of all coolness and presence of mind, as first I saw the long grey serpent-like tentacles of *Anthea cereus* waving to and fro in a clear pool? Who will restore the enthusiasm of that moment when my eye first rested on a clump of *Clavelinæ* (Plate I., fig. 4), almost as translucent as the water in which they stood! And wherefore, three weeks afterwards, could I not be induced to stop and pick up either of these, unless of very magnificent pretensions? If nature were not more inexhaustible than man's curiosity, we should come to the end of our hunting pleasures in a few years. As it is, our lifetime is too brief.

If these *first* thrills can never come back to us, there is ample compensation in the new vistas which open with increasing knowledge; the first kiss may be peculiar in its charm, but as the years roll on, we learn to love more and more the cheek on which we first found little besides that charm. Knowledge widens, and changes its horizon; and as we travel, we pass under newer skies lighted by serener stars. In direct contact with Nature we not only learn reverence by having our own insignificance forced on us, but we learn more and more to appreciate the Infinity on all sides; so that we cannot give ourselves up to one small segment of the circle, no matter how small, without speedily discerning that life piled on life would not suffice to travel over this small segment of a segment. And yet the very immensity of the world of Life is a source of encouragement. Compared with what is accessible to us, the knowledge, even of the wisest, is as that of a child; but if, instead of comparing what *is* with what is *to be known*, we compare our knowledge with our previous ignorance, the rapidity of progress becomes the keenest motive for endeavour. A few months at the coast, under proper conditions, will make us acquainted with all, or almost all, the principal forms of life; and where so much is still to be observed, each may hope to contribute something new to the general stock, and thus all be benefited.

A very few days of resolute study sufficed to substitute definite ideas for that haze which necessarily overhangs mere book-knowledge, and repeated failures helped to educate both eye and mind in the art of finding animals, and of identifying them. At first, not only did I mistake sea-weeds for polypes, but instead of filling jars and phials with ease, as anticipation had prefigured, I often came home with very meagre results—and this in a place abounding in treasures. The truth is, one has to learn many little details about the animals—where to look for them, how to *see* them when there, and how to secure them when seen—before one's basket returns home well stocked. Luck is something, of course: if there is only one bunch of sea-grapes (eggs of the cuttle-fish) thrown on shore, only one person can bag it. But it is the knowingest hunters that are the luckiest. They know how to profit by good fortune. You may perhaps be interested if I sketch a day's hunting, and into it condense most of the details, the knowledge of which may abridge your own labours, and increase your success on taking to the sport.

It is spring-tide. Little or nothing can be done during neap-tides, because it is among the rocks near extreme low-water that the prizes are found. The common smooth Anemone, and the *Daisy* indeed, may be had not far from high-water-mark in many places; and *Anthea cereus* may also be found at some distance from low-water; but for the superb *Crassicornis*, or the lovelier "Gems" and "Trogs" (*Gemmaceæ* and *Troglodytes*)—for Polypes and the rarer molluscs, we must not be far from low-water-mark. The morning is brilliant. A light breeze carries the large clouds over the lazy blue, tempering the heat of the sun; and our spirits are high, as we clatter through the tunnels on to the shore. There are three of us; and as we pass that young lady seated on a ridge, sketching Hangman's Head, she eyes us askance, and although politeness keeps in the secrecy of her own bosom the translation of that look, I know how it would run in the vernacular: "Well! how people *can* make themselves such guys!" And she who says this is, I pledge you my word, a guy of the first water. I know one when I see one, though I can't describe female costume. Her complexion is dubious, not to say spotty; and from it stands a nose not aquiline—to tell the truth, it was a turn-up,—and probably some subtle sense of harmony made her turn up very much the sides of her stone-coloured felt hat, which, with its floating ribbons and feather, may be said, in painter's phrase, to have "carried off" the nose. I also remember the three deep flounces of her Manchester muslin, and a general appearance of flying ribbons and miscellanies. If I allude to the personal appearance of this future mother of good, but not handsome, citizens, it is because her criticism of us forced us to consider from *what* pedestal of elegance we were regarded. Not that I insinuate any idea of our *not* having looked somewhat queer. Our costume was but indifferently adapted to the drawing-room, and would have obtained small suffrage on the Boulevard des Italiens, the Prater, or Pall-Mall. You shall judge. We are a lady and two men. The lady, except that she carries a landing-net, and has taken the precaution of putting on the things which "won't spoil," has nothing out of the ordinary in her costume. We are thus arrayed: a wide-awake hat; an old coat, with manifold pockets in unexpected places, over which is slung a leathern case, containing hammer, chisel, oyster-knife, and paper-knife; trousers warranted not to spoil; *over* the trousers are drawn huge worsted stockings, over which again are drawn huge leathern boots. Mine are fisherman's boots, and come a few inches over the knee. The soles are well nailed, which is of material service in preventing our slipping so much on the rocks. Now these boots, with the worsted stockings peeping above, are not, it is true, eminently æsthetic. I will not recommend them as objects for the *Journal des Modes;* but if you consider the imperfect success which will attend any hesitation as to walking into the water,—and through it, —or if you reflect on the very mitigated pleasure of feeling the water trickle into your boots,—you will at once recognise the merit of such boots as I have just described, covered with liquid india-rubber, and well greased. Never mind the inelegance: handsome is as handsome does!

In this costume we wooed the mermaids. We brought a crowbar, to turn over the heavy stones which could not otherwise be moved, but which are worth moving, because it is under such that rarities will be hidden. It is nearly half-past eleven, and the tide will have quite run out in half an hour or more. We ought to have been here a little sooner, but there is still nearly two hours' opportunity before us—so, to work! Over these jagged rocks we spring, stride, scramble, and crawl. Whew! there was an escape! This fucus is so slippery and treacherous: had it not been for the well-nailed soles, I should have perhaps tumbled into that gully,—and as the height is six or seven feet, the fall could not have been soft. The mere suggestion has made me a little nervous, and I begin to doubt whether these cockney legs were meant for such progression, and whether the right man is here in the right place,—when the shout of "Here's a *Crass!*" banishes all reflections, and in a flurry of scrambling I clamber to the spot.

"Where? Let me see it."

"In that pool."

"I see nothing."

"That is because I disturbed him. He has drawn in his tentacles, and covered himself with a coating of stones, mud, and mucus; but take off your coat, tuck up your sleeve, and you will feel him at the bottom. Got him?"

"I feel nothing but a fleshy lump of something, with small stones on it."

"That's the gentleman!"

"He seems very small."

"Now he's shut up; but if you had seen him expanded—the size of a sugar-basin—superb white tentacles round a scarlet-purple disc!"

"Well, hand me the chisel."

"There! Don't place it too near him; give him room—an inch all round. Not too hard! Never mind the toughness of the rock—clay-slate isn't made of butter; but with patience and steady blows——What! you've rapped your knuckles instead of the chisel? Well, it does fall out so sometimes. While you are hammering, I'll try elsewhere."

Accordingly I am left stretched on a sloping ledge, leaning into a pool of about a foot deep, where I have to bang away at my chisel, not in the least seeing the effect of my blows, for the crumbling of the rock has made the water the colour of a London gutter. From time to time I pull my chisel out, and feel with my hand to ascertain progress. At last a piece of the rock comes away, and I bring up the Anemone named *Crassicornis*—a very ill-favoured gentleman, to judge by his present aspect; but I throw him into the jar of sea-water, in full reliance on what he will be to-night or to-morrow. He has cost me twenty minutes' hard labour; but he was worth it.

If you are anxious to know why all this pains was taken to chisel away the rock, you may learn a curious fact—namely, that Anemones, like Achilles, are invulnerable, except in one spot. They will bear an extraordinary amount of cutting and tearing if you keep their base unlacerated. Not only have I cut off portions of them for microscopic examination, as you would cut buds off a tree, but, while I write this, there are several of the exquisite little *Auroras* and *Venustas* which have been cut or torn in half by the splitting of the stones on which they rested, and each half is as vigorous as if nothing had interfered with its integrity. In the course of some weeks no one will be able to trace in them that they have been wounded. The Abbé Dicquemare relates how he cut an Anemone in two, transversely: the upper portion at once expanded its tentacles, and began feeding; in about two months tentacles began to grow from the cut extremity of the other portion, and thus he got two perfect Anemones in place of one. And yet these animals, so indifferent to wounds, rarely survive a slight laceration of their base. At least the *Crassicornis* does not. I have not experimented in this way on the other kinds, and will limit my statement to the *Crassicornis*. This is the reason why a chisel is necessary; for the "Crass" clings to the rock with a vigour which generally defies finger-nails—unlike the *Anthea* or the *Mesembryanthemum*, which yield to a very light fingering.

I have got my prize, but have so disturbed the water that it is useless to remain longer by this pool. There are plenty more. I poke and peer into them without result, till at last a huge wall of stone rears itself in my path; and I suspect the other side is rent with fissures, rugged with ledges of promise. It is so. I squeeze into one of these fissures, where various-coloured Sponges, compound Ascidians, Serpulæ, and Algæ, with drops of water pendent from their tips, are just discernible through the darkness. In vain I strain my eyes, now familiar with obscurity; nothing tempts me. The Sponges and the *Rednoses* squirt water at me incessantly; the Algæ drip, drip, drip; sporadic Crabs trundle away in all directions; but nothing solicits my desires. You want to know why I poke into that dark hole? Because Experience—the best of schoolmasters, were not the fees so heavy!—has taught, that the two conditions most favourable to most of those marine animals we are in quest of are Darkness and Depth of water. They are impatient of the light, and prefer darkness even to many fathoms. When I say *they*, I mean most Molluscs, Crustacea, Annelids, and Zoophytes. Jellyfish seek the light, and float at the top of the sunny sea; but we shall find none of them to-day, so that fissures, caves, and the under side of boulders, must be our fields. It is well for the young hunter to bear in mind this requisite of darkness. Let him turn over all stones, peer into all fissures, push aside the overhanging Fucus, or long waving Oar-weed, and see if the pools beneath do not contain what he seeks. And when I say *look*, he must not understand thereby a careless casual glance, but a long deliberate scrutiny. He must allow the eye to rest long enough on the spot to lose the perplexity occasioned by a hundred different details, and must let "the demure travel of his regard" pass calmly over it. Sometimes the pool is so dark and still, that it is not until your nose-tip is cold against the surface that you know there is water. We have just climbed up a ledge, and looked down into a pool. Our footing is somewhat insecure, but we cling savagely, and call down few blessings on the heads of the countless *Balani* which stud the rock, and tear our hands. There, now, we have settled into a position in which we can work. Look at that *Gem*, with its lovely tiger-tentacles; it has just swallowed a small fish, and is now, while digesting, opening its arms for more. And there, on the green broad leaves of the Ulva, crawl two Sea-hares (*Aplysiæ*) (see Plate II., fig. 4). What queer creatures! One would fancy them slugs which had been troubled with absurd caprices of metamorphosis, and having first thought of passing from the form of slugs to that of hares, changed their weak minds, and resolved on being camels; but no sooner was the hump complete, than they bethought them that, after all, the highest thing in life was to be a slug—and so as slugs they finished their development. Not, however, without further caprices, since, instead of filling its mouth with teeth to grind its vegetable food, the Sea-hare transposes its teeth into its stomach, or rather into *one* of its stomachs, and not the one nearest the mouth, but the last of the series, as it passes into the intestine; so that after the food has undergone preparatory digestion, it has to be further ground by these teeth. This strange animal, as harmless as a butterfly, carries a traditional terror to the vulgar mind. The Romans believed—what would they not believe!—that the mere sight of it caused sickness, sometimes death; and pregnant women were especially warned against it. Apuleius happened to have a curiosity about this animal, for which he was accused of magic.* On every coast the fishermen who happen to know anything about the Sea-hare (and they know very little of animals they do not sell), assure you of its poisonous qualities; and the bright purple fluid it throws out, when irritated, although perfectly harmless, may well excite the suspicion of the ignorant. Whenever you find one crawling on the sea-weed, or left stranded by the retiring tide, carry it home and study it. Few molluscs are so easy to dissect; and the attention of anatomists will profitably be directed to it, because several errors are stereotyped in our treatises, which prove that, since Cuvier, few have minutely examined its structure.

But to return to our hunt. We place these Sea-hares in a small jar by themselves, and quickly add thereto a broad white ribbon of tiniest beads, which is coiled up against the under side of the ledge, and which we see with joy to be the spawn of the *Doris*—another sea-slug, if a name so ugly as that can properly be applied to a creature so attractive.—(Plate II., fig. 2.) Really this pool is enchanting! How gracefully the Polypes wave from its sides, like fairy fir-trees in the summer air. The longer we look, the more beauties

---

* Dans sa défense, Apulée répondit qu'en effet il avait observé des lièvres marins, mais seulement dans le but de satisfaire une curiosité qui n'offrait rien de condamnable. La description qu'il donne de petits osselets existans dans l'estomac de ces animaux, prouve qu'il les avait observés en naturaliste.—CUVIER, *Hist. des Sciences Naturelles*, i. 287.

and wonders we discover. I have just detected an *Ascidian** standing up like an amphora of crystal, containing strange wine of yellow and scarlet; and crawling about the root of that Oar-weed, I see various Annelids of great beauty; we must have the root—the more so that it bears some *Botryllus* clustering round it. You want to know what is that jelly-like globule no bigger than a pea? I can't answer; but probably the ovum of some fish. At any rate, the rule is to carry home whatever one does not know, and identify there, if possible; so pop the globule into a phial. Having made this haul, we may now begin to hammer away for the *Gem*. There, he is all safe in the jar, and we get down from our ledge much richer than we got up. It was a good find that pool, was it not? We had been upwards of an hour peering about, without finding anything except a *Crassicornis;* and lo! we come upon a little pool not two feet in length, which yields us enough to occupy a month of careful study.

The tide is fast flowing in, and our jars are still half empty. We must waste no time in talk. Here, give me the landing-net; I see a fish worth having. Bravo! he is in the glass jar, and looking at us with strange human look, not in the least abashed by our admiration. Did you ever see anything more exquisite? It is a ribbon-fish, but not the *Gymnetrus Banksii*. It cannot be more than two inches and a half long, and a fifth

---

* Plate I., Frontispiece, fig. 4, represents a compound Ascidian, magnified; the solitary Ascidian is less elongated, and is ordinarily of about the size figured, but is sometimes found thrice the size.

of an inch high; and note how the caudal fin, instead of being a climax to the tail, as in other fishes, forms a delicate ridge running all down the back. What a delicate Quaker brown the colour is, and how the transparency of the tissues allows us to see the pulsating heart! I hope we shall be able to keep it alive; it will be the cynosure of our collection.

Meanwhile one of our party who has been ferreting everywhere, is now crouching in a pool, and presently calls to us to come and see a *Terebella* (Plate VII., fig. 1). In three rapid strides we are there, crouch down, look where he points, and see—nothing.

"Impossible! Don't you see long waving threads, like minute worms?"

"Yes, I see threads, but that's all."

"That is the Terebella. His body is snug in the mud, and he pokes his long arms out in this way for some purpose or other, to me unknown."

"Perhaps for respiration?"

"Why do you say that?"

"Because it's *safe*. Whenever zoologists don't know the function of an appendage, they are pretty sure to say it's connected with respiration; every unknown spot is an eye, every appendage a gill, or subsidiary to gills! However, the Terebella has already been credited with branchial tufts, in the shape of smaller and redder little worms beneath the tentacles; so never mind about function*—get the animal, which I have never seen out of books."

* See the next Chapter for an elucidation of this point.

"He is hidden in the mud; we must dig out the mud."

Whereupon my companion, tucking up his sleeve, plunges his hand into the mass of sand and shells, and strews the handful on a boulder, where we soon find the worm twisting itself into irritated convolutions, as if highly disapproving of this treatment. We pop him into a phial with some sand, and he soon makes himself happy there. During this capture, quick female eyes have discerned, and nimble fingers have delicately secured, one of the loveliest of sea charmers—an *Eolis*, of about three-quarters of inch in length, with transparent body, tapering into the most graceful of tails (we must call it a tail, although anatomists call it a foot), and with rows of pink papillæ on its back, forming the most elegant of ornaments (Plate II., fig. 1). The tide may now drive in as fast as it will, we shall go home rich.

Wearied with hammering, clambering, and stooping in this blazing sunlight of a summer noon, we seat ourselves on a convenient boulder, for half an hour's repose. My companion, whose legs are lolling in a shallow pool, brings out a pocket-pistol of sherry and a bag of biscuits. To this "repast we do ample justice" (as detestable writers with unerring unanimity always say, when they want to describe eating and drinking), and then the blue lazy curl of a mild havannah rises into the warm air, making contentment more content. The waves are crawling over the boulders, and rushing up the gullies with a soothing sound. A few white sails dot the blue breadth before us. Out there on the strip of sand in the creek, a row of lazy gulls, motionless as stones, and looking like them, seem as if they too were resting from their hunt. A sense of pleasant weariness gives its dreamy calmness to the scene. We are silent, or wander into idlest chat, as if we had fairly reached that land

"Wherein it seemed always afternoon."

It was enough that our glance should fall upon the stealthy sea, and follow wave after wave as each grew out of the swell and ran along, a curling line of foam, to plunge upon the shore. We wanted nothing more. There is a peculiar charm about the sea; it is always the same, yet never monotonous. Mr Gosse has well observed, that you soon get tired of looking at the loveliest field, but never of the rolling waves. The secret, perhaps, is that the field does not *seem* alive; the sea is life-abounding. Profoundly mysterious as the field is, with its countless forms of life, the aspect does not irresistibly and at once coerce the mind to think of subjects mysterious and awful—it carries with it no ineradicable associations of terror and awe, such as are borne in every murmur of old ocean—and thus is neither so terrible nor so suggestive. As we look from the cliffs, every wave has its history; every swell keeps up suspense—will it break now, or will it melt into that larger wave? And the log which floats so aimlessly on the wave, and now is carried under again like a drowning wretch,—is it the fragment of some ship which has struck miles and miles away, far from all help and all pity, unseen except of Heaven, with no messenger of its agony to earth, except this log which floats so buoyantly on the tide? We may weave some such tragic story, as we idly watch the fluctuating advance of the dark log; but whatever we weave, the story will not be wholly tragic, for the beauty and serenity of the scene are sure to assert their influences. O mighty and unfathomable sea! O terrible familiar! O grand and mysterious passion! In thy gentleness thou art terrible, when sleep smiles on thy quiet-heaving breast; in thy wrath and thunder thou art beautiful! By the light of rising or of setting suns, in grey dawn or garish day, in twilight or in sullen storms of darkness, ever and everywhere beautiful; the poets have sung of thee, the painters have painted thee; but neither the song of the poet, nor the cunning of the painter's hand, has caught more than faint reflexes of thy incommunicable grandeur, thy loveliness inexhaustible!

During this digression our cigars have come to an end, and the tide has almost cut off our retreat. We clutch up our baskets, and with belated strides hasten over ridges out of harm's way. Our return home brings us on to the Capstone Parade, where our appearance must of course stimulate quizzing. If that young lady with the sketch-book, who saw us going out, made private reflections on the imperfect elegance of our costume, I leave you to judge of the impression we produce on the mind of that haughty "swell" with a telescope, and a mustache of recent growth. He has come to Ilfracombe with apparently no other object than that of setting his mind seriously to these things: he will array himself in a straw-hat with a pink ribbon, a coloured shirt, a shooting-jacket never meant to shoot in, and thus arrayed he will show himself and his telescope. The telescope is indispensable. He will never use it, but he borrows from it a nautical air, which is quite the right thing, you know. I wish I were just enough acquainted with that young gentleman to bow to him—I would do it in sight of the whole Parade. As we pass along, the staring excited by our incongruous appearance of dirt, damp, and utensils, suggests ludicrous reflections on the way we all judge of each other; and more serious reflections on the utterly foolish disposal of time which the majority of sea-side visitors make.

History proves that we English are a magnificent race; but I appeal to every one whether the concrete Englishman he meets abroad, or at the sea-side, in the least represents his idea of that magnificent race. I'm afraid we are disagreeable to the backbone. At the coast we are all dismal as well as disagreeable. What an air of weariness hangs over almost everybody! After the "visitors" have had their first walk on the beach, their first two or three hours' "sail," from which they return looking very green—after they have seen the sunset once, they relapse into utter novel-reading. Not only do they here read more novels than at home, but they are content to read the novels no one reads at home. Look at that young gentleman who has brought two volumes with him to the Parade. He finds the place so dull that he must read even when in the open air; yet when at home he has not the reputation of a severe student; he is not known to read at his meals, or burn the midnight oil; he is rather a stupid young gentleman, if the truth be told, and eyes us and our jars with measureless contempt, wondering "what the doose we can do, you know, with that sort of thing, you know." Then, again, I should not call that lady who scrutinises us gloomily through the blue veil of a cavernous bonnet; nor that severe and "rather intellectual-looking" lady with crimp curls, whom we presume to be a schoolmistress; nor that grim gentleman, who, we are sure, is a Methodist with an imperfect liver; nor those three sisters in their teens, "sent" to the coast with their governess,—not one of these should I call successful at the sea-side. Indeed, I meet with very few successes. The children, of course, are excluded. Master Tommy, in gorgeous hat and feathers, may "worrit" his maid by the persistence he displays in "getting hisself wet;" but the young rascal has got a spade, and means to enjoy himself, and does what he means. Another perfectly happy person amid all this weariness is yon elderly gentleman, with large stomach, white waistcoat, and a general sense of "well to do," who has escaped from care for a few days; who enjoyed his dinner at the hotel yesterday, though the port was fruity; enjoyed his breakfast this morning; and now, having read the paper, is sniffing the breezes for an appetite, and is aglow with the pleasant sensation compounded of present vigour and boyish associations. He is too old for the circulating library; has outlived straw-hats and coloured shirts; and is supremely indifferent to telescopes. He is happy. He gives a genial glance of interest to everything. He stops us, and politely inquires about the contents of our baskets, listening to the brief details with "dear me! bless me! well, how very singular!" and even thinks he should like to go out collecting himself,—if he were younger.

If the promenaders are not supremely interesting, the scene itself is worth a visit. The Capstone Parade, a walk cut round the Capstone at great expense, offers many pictures. We are at the farther end, nearest the quay, and look back upon old Hillsborough jutting out far into the water, while behind him looms the giant Hangman, grim as his name, and beyond that the purple line of another headland. Between us and Hillsborough stands Lantern Hill, a picturesque mass of green and grey, surmounted by an old bit of building which was once a convent, and which looks as if it were the habitation of some huge mollusc that had secreted its shell from the material of the rock. Indeed the houses all about naturally recall the curious shells and habitats with which our hunting has made us familiar. In mountainous districts, where houses and clusters of houses look so tiny in comparison with the huge limbs of Mother Earth, one is apt to think of man as a parasitic animal living on a grander creature—an epizoon nestling in the skin of this planetary organism, which rolls through space like a ciliated ovum rolling through a drop of water. In flat districts a town looks imposing; even a single house raises its head with haughtiness. There is nothing around to rival it in height; and from it we may fondly imagine earth our pedestal. But our thoughts are otherwise when we see the house lost on the broad side of a noble hill; and still more when, from a little distance, we see a number of houses clustered on the side, clinging to it like so many Barnacles clinging to a rock; we then begin to think of our family resemblance to all other building, burrowing, house-appropriating animals. In vain does our pride rebel at the thought of consanguinity with a mollusc; the difference between Brown, with the house he built, and Buccinum, with the shell he secreted, lies in the number of steps or phenomena interposed between the fact of individual existence and the completion of the building. Brown is aghast at the suggestion, and says he hates metaphysics. This much he will perhaps admit, namely, that whatever other advantages our habitations may have over those of insects and molluscs, it is clear they have not the advantage in architectural beauty subservient to utility. Consider man from a distance—look at him as a shell-fish—and it must be confessed that his habitation is surprisingly ugly. Only after a great many intermediate "steps or phenomena" does he contrive to secrete here and there a Palace or a Parthenon which enchants the eye.

While thus moralising we have reached our lodgings, and another work begins. Our treasures must be displayed, and, where needful, identified. The animals are to be kept alive, their wants attended to, and their habits watched, that we may form some idea of their theory of life, before we dissect them to learn something of their structure. Jars and phials are emptied into soup-plates of sea-water, previous to a general distribution into pans and vases. A glass tank is very elegant, but expensive. It is ornamental in a quite other style than that of wax-flowers, gorgons in old china, or dark specimens of the Bad Masters, which by many are supposed to enliven apartments; but if you intend to keep animals for study, a glass tank on many accounts is less desirable than several glass vases, which are inexpensive and portable. I had *no* tank, and of course never thought of transporting it to Devonshire. Up to this time my Aquarium had been constituted by finger-glasses, tumblers, and glass sugar-basins; these sufficed for the produce of fresh-water ponds; but now, on the eve of cultivating the more imposing acquaintance of marine inhabitants, I adopted a friend's advice, and laid in a store of glass jars of formidable dimensions—jars such as confectioners use to contain sponge-cakes, almonds, &c. These made an additional hamper to my luggage, and the "glass, with care," increased my anxiety not a little. I cannot enumerate the extra sixpences it cost me to impress on porters and railway guards the inherent frangibility of glass. I made myself a torment to all officials by the impressive emphasis of my anxiety. And, after all, the jars were almost worthless. Experience flatly and peremptorily decided against them, as too deep and unwieldy. I quickly discarded all but the smallest, and bought half-a-dozen glass jars of nearly a foot high, which have proved very serviceable. When an animal dies, and the mortality is great, it is easier to discover and remove the corpse, and change the water from a small jar than from a tank : moreover, in jars you can keep your animals separate ; and animals are not more amiable to each other than men ; the strong devour the weak without any religious scruples. To the jars were added shallow earthenware pans, for Actiniæ, and some animals which the Actiniæ would not molest.

Our day's produce fairly sorted, the work of identification begins. It is not enough to know that we have got a Polype, an Eolis, or an Annelid before us ; we also desire to know what species of each ; and this is sometimes a work of long and troublesome investigation, because even if the species is not one hitherto undescribed, we may have great difficulty in identifying it by descriptions. This tries the patience, but it exercises the faculties, and greatly sharpens knowledge by forcing attention upon details.

And here a word respecting the books you ought to put in your box. For reading, properly so called, the naturalist has little time while at the coast; but certain books will be constantly referred to. *All* the books on Natural History, or Comparative Anatomy, you can buy, beg, or borrow, will be found of use ; but if your portmanteau refuses the burden of many volumes, it is well you should know what will be most serviceable.

First, then, as indispensable, there must be an "Animal Kingdom"—if not Cuvier's, then Vogt's "Zoologische Briefe," or Rymer Jones's "General Outline of the Organisation of the Animal Kingdom," richly illustrated, or Mr Dallas's recently published volume, "The Natural History of the Animal Kingdom," cheap and very compact. Next you must have Mr Gosse's invaluable "Manual of Marine Zoology"—meant expressly for identification; and you may add the very cheap and compendious "Manual of the Mollusca," by Mr Woodward, published among Weale's series of Rudimentary Treatises. If you can lay hands on Johnston's "British Zoophytes," Forbes's "Naked-Eyed Medusæ" and "British Starfishes," and Alder and Hancock's "Nudibranchiate Mollusca," you will be set up. It is needless to name works of Histology or Comparative Anatomy, because, if your studies lie in these directions, you will already have possessed yourself of what is necessary.

And now, when all is done, the Microscope is taken out, and severer studies begin. The hours I spent thus, fled like minutes, and left behind them traces as of years, so crowded were they with facts new and strange, or if not absolutely new, yet new in their definiteness, and in the thoughts they suggested. The typical forms *took possession* of me. They were ever present in my waking thoughts; they filled my dreams with fantastic images; they came in troops as I lay awake during meditative morning hours; they teased me as I turned restlessly from side to side at night;

they made all things converge towards them. If I tried a little relaxation of literature, the page became a starting-point for the wandering fancy, or more obtrusive memory; a phrase like "throbbing heart" would detach my thoughts from the subject of the book, and hurry them away to the stage of the Microscope, where the heart of some embryo was pulsating. I could not look at anything intently, but the chance was that some play of light would transform itself into the image of a mollusc or a polype. THE THINGS I HAVE SEEN IN TAPIOCA PUDDING . . . !

This intense absorption in one study was wrong, and I tried to vary my employments; but intellectual passions are not obedient to abstract convictions; they will exert their jealous exclusiveness. "No array of terms can tell how much I was at ease" on matters agitating the majority of my countrymen. I utterly declined to look at the *Times*. What cared I about Palmer and his trial? or about the impending quarrel with America? As much as the stock-broker towards the close of 'Change, or the Opposition member during the vote of confidence, would care for your attempt to interest him in the "extraordinary little organ discovered this morning in the tail of a tadpole—quite unsuspected by anatomists, I assure you."

This was exclusive—say narrow, if you will. I had really interest in little but what the Scalpel and Microscope would disclose. Everything was new to me, so that every step was delightful. When I discovered what had long been known to others, the pleasure of discovery was something essentially different from that of mere learning ; and when I was fortunate enough to discover what had *not* been known before, the delight in novelty was heightened by the triumph (surely not a guilty one ?) of *amour propre*. Three months of such study were worth years of lectures and readings— although the lectures and readings were necessary preparations for the full benefit of such study. But thoughts of "benefit" are after-thoughts ;—the real incentive to work is passionate fondness for the work itself ; and I know nothing in the shape of intellectual activity which I would exchange for a long day with the Microscope. This feeling is beautifully indicated by M. Quatrefages, in that page of his *Souvenirs d'un Naturaliste* in which he describes his residence on the little archipelago of Chaussey, where none lived besides himself and a few fishermen. At night, when the songs and the disputes of the fishermen gradually lapsed into silence, and nothing could be heard but the murmurs of the sea, he sat down at his square deal table, covered with the produce of his day's hunt. There he sat, before a Microscope which opened to him the world of the infinitely minute, his pencil sketching the novel forms, his pen hastily tracing the result of his observations. And thus the night advanced, till, with fingers so benumbed that he could no longer hold the scalpel, he crept into his bed as the fishermen were leaving theirs. The passage is too long to quote, but the reader can seek it in the charming book itself, the work of a naturalist—which means, an enthusiast.

Of late years the Microscope has not only become indispensable to the scientific student, but a delight to the amateur. Nevertheless certain popular errors still deter many from its employment; and now that it is no longer the costly instrument it used to be, those errors should be combated. A very general belief of its "injuring the eyes" will be found even among microscopists. On evidence the most conclusive, I deny the accusation. My own eyes, unhappily made delicate by over-study in imprudent youth, have been employed for hours daily over the Microscope without injury or fatigue. By artificial light, indeed, I find it very trying; but by daylight—which on all accounts is the best light for work—it does not produce more fatigue than any other steadfast employment of the eye. Compared with looking at pictures, for instance, the fatigue is as nothing. Nor should any uneasiness be felt at the *muscæ volitantes*, which may be observed for the first time after using the Microscope. Few eyes are altogether without them, and it is erroneous to attribute them to the Microscope, because they may not have been previously observed. The student should early learn to keep the unoccupied eye open, not to screw it up, and distort his countenance, because the sight of other objects confuses him. In a little while he will learn to attend only to the eye looking through the Microscope; and his studies will be greatly relieved thereby.

It is further said that microscopic observation is apt to be very erroneous, and that we can see whatever we wish to see. Undoubtedly men often do see what they want to see, and what no one else can recognise. But this is not the fault of the instrument. So far from the Microscope being in itself deceptive, I maintain that it is *less* so than the unassisted eye; and for this reason: all vision is mainly *inferential;* from certain appearances certain forms are inferred; this holds of the eye as well as of the Microscope, the optical principles of which are essentially the same; but while the physical conditions are similar, the mental conditions attending vision with the assisted and the unassisted eye are different. The microscopic observer is on his guard against fallacies of interpretation which seldom suggest themselves to him when observing with the naked eye; and this critical caution makes him not only less rash in interpreting appearances, but makes him anxious to verify interpretations by other means. If the contradictions of observers be cited as a proof of the deceptive nature of the instrument, what shall we say to those manifold and persistent contradictions of anatomists using their unassisted eyes? But in truth, the controversies of microscopists have rarely turned upon simple facts of appearance, they have been almost wholly questions of interpretation.

The scientific Naturalist will not content himself with Observation, however cautious and patient; to it he will add what may be called the great mental instrument, Experiment, the instrument by which we *verify* the accuracy of our observations and conclusions, and by which Nature is interrogated. Experimental Physiology is in its infancy; which is another way of saying that the science of life is in its youth. Sciences begin in casual observation and systematic reasoning. Careless of facts, men are then careful in logic. They build elaborate structures upon shifting sand. Afterwards arrives the epoch of doubt; men become aware how illusory is the reliance on reasoning, be it never so logical, unless the data are exact, unless each step has been verified. The scrutiny of facts becoming more urgent, Observation ceases to be casual and careless; a cultivated caution takes its place, and conclusions are tested less by their logical coherence than by their verified dependence on verified facts. Now the most puissant instrument of verification is the Experimental Method which, by a process of elimination and exclusion, directly interrogates Nature.

You thus perceive that time is not likely to hang heavy on our hands, while at the coast; there is enough to do and to enjoy. Dr Johnson said that he who would acquire a pure English style must give his days and nights to Addison. I have some doubts whether the prescription is likely to be followed, or, if followed, likely to effect its purpose; but its language may be borrowed to suit my turn. He who would learn the exquisite delights Nature has for those who ardently pursue her, and would acquire a deep sense of reverence and piety in presence of the great and unfathomable mysteries which encompass Life, must give his mornings to laborious searchings on the rocks, his afternoons to patient labour with the Microscope.

CHAPTER II.

SUPERIOR CHARMS OF MARINE-HUNTING—MORNING LABOURS—THE HERMIT-CRABS—CORSICAN BROTHERS—FOOD OF THE CRAB—A COMICAL FISH—ZOOLOGICAL PARADOXES—STUDY OF NATURE—VITALITY OF SEPARATED PARTS—DEFINITION OF LIFE—ORGANIC LAW OF DIVISION OF LABOUR—THE TEREBELLA: ITS SPONTANEOUS FISSION (BONNET'S EXPERIMENTS), ITS TWO CIRCULATING FLUIDS, AND DOUBLE RESPIRATION—STRUCTURE AND FUNCTION OF ITS TENTACLES—ERRONEOUS USE OF THE TERMS MUSCLE AND NERVE—ALWAYS CARRY A BASKET—THE LANES OF ILFRACOMBE.

IN the previous chapter, I endeavoured to convey some idea of the charms which the naturalist and amateur may find in the dark fissures of frowning rocks, the endless occupation and amusement of clambering over ridges, creeping under ledges, wriggling into crevices, or exploring the under side of boulders, while a summer sun is gleaming over the retiring sea, and the white gulls are hovering almost as lazily as the whiter clouds hanging in the blue above them. Above and around, the landscape; in pools and crevices, the game; and by your side, pleasant companions eager as yourself. My description of these delights may have been thought enthusiastic by those to whom such pleasures are unknown; but in truth no enthusiasm is adequate, no description can reach the vividness of reality; the best description is but thin and meagre, following beggar-like in the footsteps of Reality. The language of enthusiasm may serve to convey to others an impression that the speaker *is* moved, but it necessarily fails to paint the felicitous details which moved him.

In this approximative and confessedly incomplete style, I will endeavour to describe something of the delights which attend the naturalist when his hunting is over, and his home is reached. For, understand this: the naturalist, and especially the physiologist, has a Morrow to his pleasure, whereas all other hunters have but a fine To-day. Far be it from me to underrate any man's pleasure; nevertheless the most catholic may discriminate; and I must here discriminate between the sportsman's possible pleasure and my own. Brown is excited when he brings down a buck, lands a pike, or recovers a snipe which has fallen among the reeds. He has his day's sport, has proved his skill—to his own satisfaction entirely proved it; and now nothing remains but to eat the produce. A dish the more upon his dinner-table—nothing but that! Not that I mean to speak disrespectfully of dishes; assuredly not of venison, pike, or snipe, well dressed, well served, well wined, and well companioned. I have no patience with those who pretend not to care for their dinner, on the ludicrous assumption that "spiritual" negations imply superior souls. A man who is careless about his dinner is generally a man of flaccid body, and of feeble mind; as old Samuel Johnson authoritatively said, "Sir, a man seldom thinks with more earnestness of anything than he does of his dinner; and if he cannot get that well dressed, *he should be suspected of inaccuracy in other things.*" *Homo sum, et nihil, &c. &c.* I respect man, and *all* his appetites. When the man is not basely insensible to the hunger of soul, the keen intellectual voracities and emotional desires, he is all the healthier, all the stronger, all the better for a noble capacity for food—a capacity which becomes noble when it ministers to a fine, and not merely a gluttonous nature. Moreover, I observe this constant fact, which is worth flinging at the heads of all super-refined superfine spiritualists, who talk about our God-given senses as "gross"—namely, that whenever we get authentic details about a great man, we always find him to have been a generous eater. If I, who write this, must confess to being a small eater, I must also confess to not being a great man. Had nature willed it otherwise . . . . . but she did *not* so will it; and only gave me sufficient sagacity to perceive that dishes are in no sense despicable.

When, therefore, I think of the hunter's finale as merely an extra dish, and pronounce that to be an anticlimax to his day's work, instead of being, as my finale is, an ascending crescending culmination of delight; this reflection is not suggested by any scorn of eating in itself, but is suggested by the obtrusive fact, that eating is at the best a *finite* pleasure. It has no savour of the infinite, which all true and great pleasures must possess. It is vigorous in sensation, but it is circumscribed; it throws out no feelers into other, wider regions; it generates no thought; it leads nowhither; it is terminal. Therefore, I say the finale of the table is an anticlimax for a hunter; unless, indeed, he is hunting for subsistence, and then of course his finale becomes proportionately aggrandised.

No such anticlimax was mine; no such terminal enjoyment; my finale was not final. If, as a matter of fact, the dissecting-table was the scene on which my captures made a last appearance, this last appearance was the end of a long series of episodes intermediate between the capture of prey and the incision of the scalpel. And even this finale was not, strictly speaking, a *finis;* for when the last shred of delicate tissue had been examined under the Microscope, when various parts of the animal had been made into "preparations" for after-study, when everything to the physical eye may have seemed concluded, no *end* was reached, no dead wall of terminal blankness; on the contrary, the philosophical eye followed the devious paths of speculation, into which new facts conducted; and thus the feast of reason and the flow of physiology, generated pleasures superior to the pleasures of the ordinary hunter by quite transcendent degrees. I dined as well as Brown, thanks to my poulterer and fishmonger. If the truth were known, my game was perhaps better than his. We both dined,

"But oh! the difference to me!"

On an equality as regards mere plenitude and digestive beatitude, how far below the "reaches of my soul" were any thoughts which he could extricate from under that oppression of venison!

Table for Table, then, finale for finale, it is clear that my hunting was superior to Brown's in having a grand climax; but I had *already* distanced him by many lengths before we came to that winning-post of the table. Brown lands his pike, and carries it home with a careless ostentation, and an "Oh-I-could-have-caught-more" kind of air. Admiring eyes follow him through the village. He stands on his lawn and holds up the fish before the window, to receive the facile admiration of acquiescent Mrs B. And *here* his sport ends till dinner. Now, although I carry home a basket of marine animals with none of that effect upon the popular mind (indeed the popular mind is terribly apt to eye my costume and basket with ill-concealed contempt), and although my servant can't for her life think what master does with them things, not she; yet, when Brown and I are both fairly housed, his delight runs down like a clock-weight—mine ascends like a windlass. The amusement of distributing and identifying the animals I have already noticed; so we will suppose all that over, and that the fatigues of the day have been snored off with great frankness. The morrow begins.

My first thought on descending in the morning is to glance with fond anxiety at my animals. While the urn is musically hissing, and the coffee percolating, I am carefully inspecting vases and pans, removing a bit of dirt here, a decayed weed there, placing a small stone more conveniently there, poking a sluggish *Doris* (Plate II., fig. 2), to assure myself that he is alive, rescuing an *Actinia* from the crowding propensities of its cousins, or — sadder office still ! — discovering and removing those of my pets who have been inconsiderate enough to pay their debt to nature's laws. This removal is a very necessary bit of work, for these amiable creatures, when dead, are capable of stinking with some vigour, and corrupting the water in which their companions live. Breakfast was always ready before I had fairly finished my overseeing, for the parishes were numerous, and some of the parishioners apt to skulk out of sight. During the pleasant hour of breakfast, and the cigar which followed, I contemplated my treasures with a placid eye. Picture to yourself a large and airy room, made out of two, in an elegant villa: on the sideboard stand four or five glass vases, various in size and in contents ; from this the eye travels to a table, opposite the window which opens on a balcony sheltered by a verandah ; this table is covered with bottles, phials, troughs, microscope, dissecting-case, note-book, &c., all in that state of imperfect order denominated higgledy-piggledy. Three soup-plates occupy the extreme end of the table, and, "carry the eye" into the balcony, where three yellow earthenware pans and a white foot-pan mimic, *tant bien que mal*, the shallow rock-pools of the shore. If the eye so carried into the balcony happen to be in the least a conventional eye — one never so well pleased as when resting on the elegancies of surface civilisation — it is possible, nay, it is extremely probable, that it will rest upon these pans with a very mitigated admiration. Even I will confess they are not strictly ornamental. Without having read "Price on the Picturesque," one may be startled, on walking up an elegant garden to an elegant villa, if the eye falls upon yellow pie-dishes and white foot-pans symmetrically insolent under a verandah. As to Gillow of Oxford Street, be sure he would feel his hair turning grey at such a sight. And I know many persons of irreproachable drawing-rooms, liberal in opinions, affable in demeanour, and glad to own me among their visiting acquaintances, who would cut me at once after seeing the proprieties thus outraged. But look inside my pans and pie-dishes, and if you are not pleased with the beauty of those exquisite animals, and those charming weeds, I set you down as one who judges of books by their binding, not by their contents. Observe, I do not take my stand on these pie-dishes. I should greatly prefer a tank, either of glass or stone; but one can't improvise a tank at sea-side lodgings, whereas pie-dishes *are* attainable.

From the glimpse just given of my before-breakfast occupation, you begin to suspect something of what was meant by the intermediate episodes between the capture and the scalpel; you see that the mere keeping and watching of these animals will be a source of pleasures unattempted yet in prose or rhyme. One gets interested in anything which solicits attention, and in proportion to the solicitation; hence our fondness for animals and children. Nay, do but watch a man walking round his garden, pulling out this weed and brushing off that insect, trimming this branch and trailing up that cluster,—see what an incessant pleasure it is to him. Now deepen this pleasure by a scientific interest, which makes every detail of manners, and every newly-observed point of structure, the starting-point for fresh speculation, and you will form a faint idea of what it is to keep pans and vases full of animals.

You doubtless know the Hermit-crab, by naturalists named *Pagurus?* Unlike other crabs, who are content to live in their own solid shells, Pagurus lives in the empty shell of some mollusc. He looks fiercely upon the world from out of this apparently inconvenient tub, the Diogenes of Crustacea, and wears an expression of conscious yet defiant theft, as if he knew the rightful owner of the shell, or his relatives, were coming every moment to recover it, and he, for his part, very much wished they might get it. All the fore part of Pagurus, including his claws, is defended by the solid armour of crabs. But his hind parts are soft, covered only by a delicate membrane, in which the anatomist, however, detects shell-plates in a rudimentary condition. Now a gentleman so extremely pugnacious, troubled with so tender a back and continuation, would fare ill in this combative world, had he not some means of redressing the wrong done him at birth; accordingly he selects an empty shell of convenient size, into which he pops his tender tail, fastening on by the hooks on each side of his tail; and having thus secured his rear, he scuttles over the sea-bed, a grotesque but philosophic marauder. You ask how it is that this tendency to inhabit the shells of molluscs became organised in the hermit-crab? Either we must suppose that the crab was originally so created,—designed with the express view of inhabiting shells, to which end his structure was arranged; or— and this I think the more reasonable supposition— that the hermit-crab originally was furnished with shell-plates for the hinder part of his body, but that these have now become rudimentary, in consequence of the animal's practice of inhabiting other shells,— a practice originally resorted to, perhaps, as a refuge from more powerful enemies, and now become an organised tendency in the species.

Be this as it may, the hermit-crab will not live long out of an appropriated shell; and very ludicrous was the scene I witnessed between two taken from their shells. Selecting them nearly equal in size, I dropped them, "naked as their mother bore them," into a glass vase of sea water. They did not seem comfortable, and carefully avoided each other. I then placed one of the empty shells (first breaking off its spiral point) between them, and at once the contest commenced. One made direct for the shell, poked into it an inquiring claw, and having satisfied his cautious mind that all was safe, slipped in his tail with ludicrous agility, and, fastening on by his hooks, scuttled away, rejoicing. He was not left long in undisturbed possession.

His rival approached with strictly dishonourable intentions; and they both walked round and round the vase, eyeing each other with settled malignity,—like Charles Kean and Wigan in the famous duel of the *Corsican Brothers*. No words of mine can describe our shouts of laughter at this ludicrous combat,—one combatant uneasy about his unprotected rear, the other sublimely awkward in his borrowed armour. For the sake of distinctness, I will take a liberty with two actors' names, and continue to designate our two crabs as Charles Kean and Alfred Wigan. C. K., although the blacker, larger, and stronger of the two, was at the disadvantage of being out of his shell, and was slow in coming to close quarters; at last, after many hesitations, approaches, and retreats, he made a rush behind, seized the shell in his powerful grasp, while with his huge claw he haled Wigan out, flung him discomfited aside, and popped his tail into the shell. Wigan looked piteous for a few moments, but soon, his "soul in arms and eager for the shell," he rushed upon his foe; and then came the tug of crabs! C. K. had too firm a hold; he could not be dislodged. I poked his tender tail, which was exposed through the broken shell, and he vacated, leaving Wigan once more in possession. But not for long. Once more Wigan was clutched, haled out, and flung away. I then placed a smaller shell, but perfect, in the vase. Kean at once quitted his dilapidated roof, and ensconced himself in this more modest cottage, leaving Wigan to make himself comfortable in the ruin; which he did.

The fun was not over yet. A third hermit-crab was placed in the vase. He was much smaller than the other two, but his shell was larger than the one in which Kean had settled, as that unscrupulous crab quickly perceived, for he set about bullying the stranger, who, however, had a shell large enough to admit his whole body, and into it he withdrew. It was droll to see Kean clutching the shell, vainly waiting for the stranger to protrude enough of his body to permit of a good grasp and a tug; but the stranger knew better. He must have been worn out at last, however, for although I did not witness the feat, an hour afterwards I saw Kean comfortable in the stranger's house. They were changed again; but again the usurpation was successful. On the third day I find recorded in my journal: "The crabs have been fighting, and changing their abodes continually. C. K. is the terror of the other two, and Wigan is so subdued by constant defeats that he is thrown into a fluster if even an empty shell is placed near him; and although without a shell himself, which must make him very cold and comfortless in the terminal regions, he is afraid to enter an empty one. The terrors of the last two days have been too much for his nerves: one must almost question his perfect sanity; he is not only beside his shell, but beside himself. The approach of C. K. throws him into a trepidation, which expresses itself in the most grotesque efforts at escape."

A new experiment was tried. Throwing a good-sized whelk into the vase, I waited to see Kean devour the whelk in order to appropriate his shell; for the house he last stole, though better than the previous houses, by no means suited him. Mr Bell, in his *History of British Crustacea*, conjectures that the hermit-crab often eats the mollusc in whose shell he is found—a conjecture adopted by subsequent writers, although Mr Bell owns that he never witnessed the fact. My observation flatly contradicted the conjecture. Kean clutched the shell at once, and poked in his interrogatory claw, which, touching the operculum of the whelk, made that animal withdraw and leave an empty space, into which Kean popped his tail. In a few minutes the whelk, tired of this confinement in his own house, and all alarm being over, began to protrude himself, and in doing so gently pushed C. K. before him. In vain did the intruder, feeling himself slipping, cling fiercely to the shell; with slow but irresistible pressure the mollusc ejected him. This was repeated several times, till at length C. K. gave up in despair, and contented himself with his former shell. Thus, instead of *eating* the whelk (which, I may remark in passing, the crab never does, even in captivity, where food is scanty), he had not even the means of getting him out of his shell, and the conjecture of our admirable naturalist must be erased from all Hand-books.

These traits of manners and morals pleasantly vary our graver observations: but it is only with the higher organisms that we can be so amused; the lower organisms, although they have their manners and their morals, are too far removed from us to be intelligible. I have no doubt the mollusc is a moral individual, but you cannot consider him to be greatly impassioned; an oyster, or a limpet, may have his theory of life: but you cannot appeal to his finer sensibilities through the medium of music, poetry, or painting. I have some doubts even of the crab in these regions of culture; but if he cannot soar so high as Art, we see how he touches the confines of Wit by his feeling for the Grotesque. Fish, too, are funny, and far more educable than people suppose. Your fish has a sense of the proprieties; he will even condescend to conventionalism in costume. At least, one I had at Ilfracombe did so. A queer little dolphin-like fellow he was, who, after swimming about the vase for some time, would sink to the bottom, and there, curling his tail round him as a cat does when making herself comfortable, he would look up with his impudent unabashed eyes, and pant away, as if fatigued with his gambols. This curling of himself whenever at rest was very comical, and he looked as if he knew it. When I had him, he was in full black—evening costume; but on descending next morning I found him arrayed in an entire suit of light brown—cool morning summer costume; in the afternoon he again presented himself in full black; and the next morning he was dead. I grieved for him, and, as a consolation—dissected him.

This was my constant solace, when I found—as, alas! I often found—that some of my pets had departed. The zoologist softened, the anatomist was resolved. I had lost a pet, and gained a "preparation." Grief gave way under the scalpel. Science dried afflicted eyes. Nay, shall I confess it? Many a time I have had the unfeelingness to eye a pet with an undertaker's glance, almost wishing it would die, for the sake of its corpse. And when this was the case, you may be sure I bore the announcement of mortality with something of that fortitude displayed by legatees when a choleric old gentleman, or a lady of starched and vigorous virtue, departs this life, leaving a trifle in the 3 per cents.

Death was no finale to me. The closing scene was only the close of an act, after which the curtain rose once more, the drama culminating in interest. A thousand problems assailed the mind; a thousand strange thoughts arose as I penetrated deeper and deeper into the mysteries of these various organisms, and mused upon their many paradoxes. Here was an animal without a heart; there, one without a liver —nothing but quantities of hepatic cells distributed along the course of the alimentary canal. Here was an animal breathing by means of his legs; and here one not breathing at all. Here was a mollusc with its intestine passing *through* its heart; here another with teeth in its stomach; and here an animal (the *Physalia*) digesting its food before swallowing it— that is to say, performing the act of chymification *before* the act of deglutition. Here was an animal of two sexes, and here one of no sex at all,—or, more correctly speaking, of the female sex only, the male being non-existent. But the most piquant of all paradoxes is that of the parasitic Crustacean, a *Lernœa* (see Plate V., fig. 1): The female, ensconced in the eye, or gills, of a fish, lives a lazy life at the fish's expense, and the male lives upon her as she lives on the fish (not unlike some disreputable males of the human species), and this male is himself infested with parasitic Vorticellæ, so that we find parasites of parasites of parasites!*

> "Great fleas have little fleas, and lesser fleas to bite 'em,
> And these again have other fleas, and so *ad infinitum*."

Paradoxes like these—and they might be indefinitely multiplied—titillate curiosity, but they do not form the real attraction of our studies; they excite a smile, or a passing wonderment, which is as nothing compared with the deep, abiding, almost awful sense of the mystery and marvel of Nature. The crowning glory is the knowledge which ever opens into newer and newer vistas, quickening our sense of the vastness and the complexity of Life. For it is eminently the case with these studies, that they intensify and exalt our conceptions of the incommunicable grandeur and infinity of Nature. Many eloquent pages have set forth the effect produced upon the mind by the study of Nature, the enlarging influences of contact with

---

* See the second part of Nordmann's *Mikrographische Beiträge zur Naturgeschichte d. Wirbellosen Thiere* 1832, for a full and admirable monograph of these parasites, illustrated by coloured plates; also a *Mémoire* by Van Beneden, in the *Annales des Sciences*, 1851.

and contemplation of her phenomena, so different from the fleeting fashions and miserable pretexts of much that passes as civilisation, so full of rebukes to our foolish pride and pretences; so full of lessons to us to be in earnest, and to trust in simple earnestness. But although contact with all reality must necessarily have something of this influence, I should say, speaking from my own experience, that this is true in quite another sense to those minds familiarised with the phenomena of life manifested by the simpler organisms. Here the Microscope is not the mere extension of a faculty, it is a new sense. At some distance from the Alps, we discern their masses of purple grandeur, but that is all we discern; on approaching nearer, these purple masses assume shapes more and more definite, although their varied architecture is still hidden from us: we see none of their ravines and valleys; a little nearer, and we detect these, but discern none of the chalets nestled in the valley, or scattered over the mountain-sides; nearer still, we see the habitations, and the cattle, and the men; yet nearer, and we discriminate individualities; but we have still to advance, and patiently watch, before the tragedies and comedies acted in these scenes can become intelligible to us. Thus with each step we have changed our conceptions of the Alps. Thus with each step do we change our conceptions of Nature. We all begin, where most of us end, with seeing things removed from us—kept distant by ignorance and the still more obscuring screen of familiarity. We afterwards learn to observe something besides these broad general outlines which constitute the scenery of our existence, and learn to admire the magnificence of Nature. The observation of one detail is a step to the recognition of many. In this stage we resemble the traveller who has discovered the Alps to have valleys and habitations. If the Microscope be now placed in our hands, it brings us into the very homes and haunts of Life; and finally, the high creative combining faculty, moving amid these novel observations, reveals something of the great drama which is incessantly enacted in every drop of water, on every inch of earth. Then, and only then, do we realise the mighty complexity, the infinite splendour of Nature; then, and only then, do we feel how full of Life, varied, intricate, marvellous, world within world, yet nowhere without space to move, is this single planet, on the crust of which we stand, and look out into shoreless space, peopled by myriads of other planets, larger, if not more wonderful, than ours. And if with this substitution of definite and particular ideas for the vague generalities with which at first we represented Nature—if with increase of knowledge there comes, as necessarily there must come, increase of reverence, it is evident that the study of Life must of all studies best nourish the mind with true philosophy.

The facts are the least of the attractions in this study, although they are the bricks with which you build. If you happen to be of a speculative turn, every fresh observation will start new trains of thought.

Walk up to my working-table, and take the first phial or trough chance may present. You have chosen a phial in which a quantity of thread-like worms are wriggling like uninspired Pythonesses. You are mistaken in supposing them to be worms,—they are nothing of the kind; they are not even individuals. In spite of your stare, I repeat the statement: they are not individuals, they are organs. Why then do they live and wriggle? and of *what* are they organs? The first question is easier to ask than to answer. The second is as easy to answer as to ask; so, like an adroit teacher, who evades difficulties to drop with confidence on what is easy, I will answer it. In the preceding chapter (p. 26) I recorded with some minuteness the finding of a *Terebella* buried in the sand, its long thread-like tentacles waving in the air being all that was visible, until it was dug up. Those tentacles are what you have in the phial before you. While examining the worm, I observed that one of its tentacles had been torn off, and was wriggling with independent vivacity: I bethought me of trying how long these organs would live separated from the body; so, cutting them all off, I placed them in this phial. This was on the 21st May; on the 25th some died; but to-day is the 27th, and there are still several vivacious.

Nor is this by any means a solitary instance. The other day I was examining one of those white filaments which certain *Actiniæ* copiously throw out when disturbed. The filament is nothing but a delicate membranous tube covered with vibratile cilia, and enclosing, I believe, a still more delicate tube, filled with granules and those thread-capsules which anatomists declare (erroneously, as I shall prove hereafter) to be the urticating or stinging cells. Such is the structure of this filament, which, although it had long been removed from the animal, was twisting and twirling itself like a worm in an unhappy state of mind; and moving across the stage with motions which it was impossible to distinguish from voluntary motions. I then crushed it into many minute fragments; but long afterwards I observed some of these moving about like so many animalcules. Another day I observed what seemed a tiny white annelid crawling at the bottom of a vase; on securing it, I found it was one of the Actinia's filaments.

It may be answered that this motion was not life: in both cases it was only ciliary action. But do not let us cheat ourselves with phrases. What is the motion of early embryos but ciliary action? and what is explained by the phrase? Where then does life begin? In that foot-pan you see a dozen lovely *Medusæ*, swimming to and fro with their laborious pulsating movements — are not those movements the finger-posts of vitality? Well, then, now attend: yesterday I was dissecting some of these, and, while examining the exquisite fringe of tentacles which hangs from the border of the disc, I observed every one of these polyp-like tentacles move to and fro, now protruded, now withdrawn into the substance of the disc, each with independent action, and this on a portion of the disc which had been many hours separated from the animal. The fringe does no more when the animal is vigorous on the warm surface of the tranquil sea; it does no less now that the animal is in shreds. Look in that saucer, and you will observe the fragment of another *Medusa;* the animal is dead, and almost melted away. I have already cut out two of the ovarial chambers, yet you see the oval tentacles are twisting about as if seeking prey. *This* is not ciliarity, but contractility. This is life, unless you restrict the term life to the meaning it carries in its highest formula. If it is motion, it is *vital* motion.

Can motion, alone, be taken as the index of life? Certainly not. But let us try to be precise in our language. Life is a complex term, indicating complex phenomena. In its highest formula, expressing all the requisite generality of what is included in the term, it indicates the triple unity of Nutrition, Reproduction, and Decay. An animal grows, reproduces, and dies; these are the three capital and cardinal facts of its organism. Out of these issue many derivative and secondary phenomena, one of which is Motion. In some animals, motion can scarcely be said to have any existence. The *Ascidians*, for example, although of rather complex structure, have nothing which approaches it, unless we should so designate the occasional contraction and dilatation of their two orifices. We may therefore conceive Life without Motion, and Motion without Life; and thus, with some plausibility, ask whether the movements exhibited by the tentacles of the *Terebella* and *Medusa* ought to be received as indications of life ? Here I get myself into a fix. The thought arises that what I observe in these tentacles is owing to a surplus residue of vitality, retained by them, not to any central source of self-renewing vitality, such as the organism possesses; consequently, inasmuch as these tentacles neither grow, nor reproduce themselves, they fail to fulfil the primary conditions of Life; in other words, they are *not* alive, in spite of movements, apparently spontaneous, during a whole week of independent existence.

In arguing with one's-self, one has always a respectful antagonist, to whose objections every attention is given. Having given due attention to myself, I now turn round upon myself, and remark with some emphasis: Very true; but you overlook the important fact that in speaking of Life as the triple unity of Nutrition, Reproduction, and Decay, you are speaking of the *whole organism;* whereas in the tentacles of *Terebella* and *Medusa*, we were considering an organ, not an organism; and to apply your definition to an organ, would be to deny its vitality altogether. The *animal* cannot be considered as wanting in either of the triple terms; but the very essence of an *organ* is that it *specialises a function*—that is to say, takes upon itself to do something for the benefit of the whole animal, in return for which it is absolved from doing many things which the animal must do. In the earliest forms of Life, as in the earliest states of Society, all do everything, each does all. There is no separate digestive system, no separate respiratory system, no muscular system, no nervous system. Every part of the animal assimilates, respires, contracts, moves; just as in barbarian tribes every man is his own tailor, his own purveyor, his own architect, and his own lawyer. At last the principle of Division of Labour emerges; then that which is true of the whole organism ceases to be true of an organ; and we have no more right to demand that an arm should digest food, than that Moses & Son should preside over the deliberations of Downing Street, or cook the Whitebait dinner; we have no more right to ask the lungs to produce offspring, than to ask Mr Cobden to take command of the Baltic fleet, and Mr Bright to perform the operation for stone. Each no longer does all. When, therefore, we look at these arms of the *Terebella*, which wriggle after a week's separation from the body, we see them manifest as much of life as they manifested a week since. They would grow if they had food; unhappily they have lost the power of preparing food, and they die at length from starvation.

But put down that phial, and look at this which contains another and far more beautiful species of Terebella, by name *Nebulosa* (Plate VII., fig. 1.) It makes itself a solid tube of earth, which it cements by a mucus exuded from its surface; and in this tube, but not attached to it, the Terebella lives, merely putting forth its long tentacles into the water. I have taken it from its tube to watch its beauty and its manners. Professor Rymer Jones, in the last edition of his *Animal*

Kingdom, says, that "our knowledge of these animals has been until recently very limited;" and he adds, "the zootomist who should enjoy favourable opportunities of inspecting the larger species in a fresh state, could hardly make a more valuable contribution to our science than by giving an account of the organisation of these interesting animals." My opportunities of observing the larger species have been null; but having dredged up many of the smaller species off Tenby and Caldy, I studied those with great eagerness; and although my observations had, for the most part, already been included in the more elaborate investigations of Milne Edwards and Dr Williams, I have yet something new to offer, little though it be.

No one, I believe, has yet recorded the fact of the Terebella multiplying itself by the process of gemmation, which is known to occur in the case of some other Annelids—such as the *Naïs* (Plate IV., fig. 3), the *Syllis*, and the *Myriana*, and of which the reader will find all the details in the accessible works cited below.* When the animal reproduces by this budding process, it begins to form a second head near the extremity of its body. After this head, other segments are in turn developed, the tail, or final segment, being the identical tail of the mother, but pushed forward by the young segments, and now belonging to the child, and only vicariously to the mother. In this state we have two

---

* QUATREFAGES: *Annales des Sciences*, 3me série, i. 22; *Souvenirs d'un Naturaliste*, i. 247. RYMER JONES: *Animal Kingdom*. OWEN: *Comparative Anatomy*, vol. i.

worms and one tail. It is as if a head were suddenly to be developed out of your lumbar vertebræ, yet still remain attached to the column, and thus produce a double-headed monster, more fantastic than fable. Or suppose you were to cut a caterpillar in half, fashion a head for the tail half, and then fasten this head to the cut end of the other half—this would give you an image of the Syllis budding. But in some worms the process does not stop here. What the mother did, the child does, and you may see at last six worms forming one continuous line, with only one tail for the six. The tail indeed is the family inheritance; but reversing the laws of primogeniture, it always descends to the youngest: like that elaborate display of baby linen which was worked with such fondness for the first-born, and has become in turns the costume of successive pledges, as they appeared on this scene of life with a constant diminendo of interest in all but parental eyes. Such, in a few words, is the budding of Annelids. I omit differences, and many curious details, only desiring to fix the reader's attention on the cardinal fact. The separation finally takes place, and then we perceive the children and grandchildren are not quite the same as their ancestor. The fact has not been observed at all hitherto in the group of Annelids named *Tubicola;* yet two of my *Terebellæ* gave me a sight of it. The first died before the separation took place. The second, after a day or two's captivity, separated itself from its appendix of a baby, and seemed all the livelier for the loss of a juvenile which had been literally in that condition of "hanging to its mother's tail," which I have heard applied in metaphorical sarcasm to small boys anxious to be with their mothers. The young one only lived four days.

This spontaneous self-division of worms is a curious fact, first observed and accurately described by the Genevese naturalist, Charles Bonnet,* whose discoveries in reproduction we shall have to consider in a subsequent chapter. He found the fresh-water Naïs frequently double, and saw the separation take place. He also cut the worm into several pieces, and observed each piece reproduce its head, and grow into a perfect worm. One observation he made is worth repeating here.† He cut a Naïs transversely, but not entirely in two; that is to say, the two portions were held together by a mere thread; in less than an hour afterwards the two were again perfectly united, and no trace remained of the operation, except a slight constriction, and an interruption in the continuity of the vessels and viscera. This interruption was very similar to such as he had observed in several worms in their ordinary condition; and if we compare the observation with that made by M. Peltier,‡ we shall see that they confirm each other. "In those species," says M. Peltier, "which have a dorsal contractile vessel, in which the course of the nutritive fluid can be traced, we see that in proportion as the fluid becomes poorer, the contraction becomes feebler, and is arrested at the

---

* BONNET: *Traité d'Insectologie*, 1745, vol. ii.   † *Ibid.* ii. 133.
‡ See *Comptes Rendus des Séances de l'Académie*, 1844, Jan. 22. p. 161.

point where the fluid, having been already absorbed, ceases to arrive. Here we see in the middle of the body, at the very point where the nutritive fluid is arrested, a constriction of the dorsal vessel, and two large absorbing vesicles are formed, which collect, for the benefit of the posterior half of the animal, the aliment no longer furnished by the anterior half. As soon as these vesicles begin to act, the posterior half of the dorsal vessel reassumes its contractile movements. These contractions have their origin in the vesicles, and have no connection with the anterior half, nor are they synchronous in their movements with it. In front of these vesicles a constriction takes place which finishes by separating the two portions into two distinct animals. If one of these be kept alive in the same drop of water for some days, the aliment gradually diminishing, the process is repeated, and this again may be repeated, so that six distinct individuals are produced from one, solely by the deficiency of food, which causes a constriction of the dorsal vessel." What an argument in favour of those who maintain starvation to be a cause of over-population! Quatrefages has noticed an analogous case in the *Synapta*. "Hunger is the sole cause of the spontaneous amputations," he says. "It would incline one to say that the animal, feeling himself unable to find food for his whole body, successively suppresses those parts which cost too much."*

It is right to add, that a very great authority on all questions relating to the Annelids—Dr Thomas Williams

---

* QUATREFAGES, *Souvenirs d'un Naturaliste*, i. 62.

of Swansea — emphatically denies the whole of the "fables" originated by Bonnet, and accepted by all successors. He says* there is not one word of truth in the statements, and he laughs at Professor Owen for repeating them. I have examined the point, as regards the Naïs, since reading his denial, and am at a loss to understand how that denial could have been made in terms so sweeping. In the *Naïs proboscidea* the spontaneous separation is preceded by the formation of a head, with the unmistakable proboscis ; as any one may observe who will collect a few specimens from a neighbouring pond.

The *Naïs proboscidea* is a convincing instance, because the long proboscis may easily be seen projecting from the segment where the separation will take place ; and its presence removes all possible doubt as to the formation of a new head. Disposed as I am to allow due weight to the opinions of Dr Williams, I am surprised to find him saying : "The tail-fragment *never*, as can be proved by easy observation, produces a single new ring or segment of the body. If this be true, how completely improbable must be the statement, that the headless piece is capable of constructing a *new head!*" No one can read Bonnet's own account of his observations and experiments, recorded with great minuteness and precision, without feeling considerable surprise at such a remark. If Dr Williams were correct, his contradiction would throw a doubt on the observations of

---

* WILLIAMS—*Report on British Annelida*, 1851 — In Reports of British Association, 1852, p. 247.

any naturalist; because if Bonnet is not to be trusted, on a point so easy of verification, when his statements are so precise, and his observations so numerous and minute, no one can be trusted. Fortunately, however, Bonnet is rigorously correct. I have verified his observations under the impulse of Dr Williams's denial; and at the very moment of writing this, I have two Naïds in separate vessels, who have reproduced their heads and proboscrs under the following circumstances. The two worms were first cut in half, the fragments which bore the heads were thrown away, and those which bore the tails were placed in vessels, with nothing but water and a little mud, in which, as I had scrupulously ascertained, no worm or other visible animal was concealed. In a few days the complete heads were formed; the heads were examined daily during their formation. When the animals were quite perfect I once more cut them in two, threw away the head-fragments, and replaced the tail-fragments in their vessels. A second time the heads were formed. A third time the experiment was repeated; and the worms are now lively, after their fourth section, so that before this passage is printed, I have little doubt they will present the fourth reproduction of head and proboscis.*

Amateurs are not fond of worms; nor, until they have seen *Serpulæ*, *Sabellæ*, and *Terebellæ*, expanding and waving their beautiful tentacles in the water, can they understand why we should take so much trouble to

---

* One died a few days afterwards, but the other made itself a new head as before; it died, however, after a fifth division.

secure them. And yet, apart from their beauty, the worms deserve our study. Their structure is full of interest.

Let us for a moment consider their blood. That some animals have red blood, and others blood not red (which made Aristotle say that some have blood and others none at all), you know perfectly well; but that the worms have blood of various colours, is probably news to you. Swammerdamm* was the first who broke down the Aristotelian division, by showing that the blood of the earthworm was red; and Cuvier extended this observation to a whole class of worms, to which he gave the name of *Vers à sang rouge;* but this was vehemently criticised by de Blainville; and recent researches, especially those of Milne Edwards, Quatrefages and Williams, have shown that a great diversity in colour exists. Thus the Sea-mouse (*Aphrodita*) has colourless blood; the *Polynoë* pale yellow; the *Sabella* olive green; and one species of *Sabella*, dark red. But this difference of colour is trifling compared with the *absence of corpuscles* from the blood of all Annelids. The corpuscles, as you know, are the floating solids of the blood, and on them devolve the most important physiological functions; but the blood of all Annelids is entirely destitute of them; and Milne Edwards, in noticing the fact, remarks that this liquid resembles the imperfect blood of the vertebrate embryo in the early periods of development.†

---

* SWAMMERDAMM: *Biblia Naturæ*, i. 119.

† MILNE EDWARDS: *Leçons sur la Phys. et l'Anat. Comparée*, 1857; vol. i. p. 107.

There is one remark I wish to make in passing, respecting the colourless blood of the majority of invertebrate animals, and that is the proof it affords of the error, not uncommon, in attributing the colouring matter of animals to the colouring matter of their blood. It is now known that the colour of the muscles is due to a peculiar pigment, far more than to the blood which is in them. It is quite clear that the purple fluid ejected by the Sea-hare, or the inky fluid ejected by the Cuttle-fish, cannot derive their colour from colourless blood. There are muscles in several Molluscs—for instance, those of the œsophagus in the Slug and Water-snail — which are of reddish hue, yet the blood of these animals has no colour. And, as a final argument, the integument of the Anemones is richly coloured, yet they have no blood at all.

To return to our Annelids. If we grant that the fluid hitherto universally regarded as blood is truly blood, we shall have to acknowledge that these Annelids have two different kinds of blood; for over and above the fluid which we see circulating in the vessels, there is a fluid circulating, or, more correctly speaking, *oscillating*, in the general cavity of the body, and *this* fluid carries with it what are called the blood-corpuscles. It consists of albumen and sea water; and is by Dr Williams named the "Chylaqueous fluid," the simplest form in which blood makes its appearance, distinguished from the "Blood-proper," in not being a fluid circulating in a system of closed vessels, but a fluid which carries the chyle directly to the tissues. An image may render the mechanism intelligible. Suppose a worm suspended in a phial of water. Let the worm represent the intestinal canal, and the glass phial represent the external integument, the water will then represent the chylaqueous fluid, which moves with every motion of the intestine, and fills up every cavity made by its motions. The albuminous and corpuscular nature of the chylaqueous fluid prove it to be subservient to the purposes of nutrition. It also serves another purpose, acting as an internal skeleton. You are surprised at the idea of a *liquid skeleton?* yet you sit on a *cushion of air*, without the least astonishment. The fact is positive: the Annelid employs its chylaqueous fluid as a fulcrum by which it moves; let the fluid out, and all power of locomotion vanishes.

The two bloods have two methods of aeration. The "chylaqueous fluid" rushes into the lovely tentacles which in many species wave above the head, and there is aerated, aided by the action of the cilia which line the inner surface of the tentacles. The "blood" is carried to those arborescent tufts without cilia, which branch from each side of the head beneath the tentacles. But although the respiratory process does undoubtedly take place in these organs, yet in animals so simply constructed, each organ performs more than one function. Let us hear Dr Williams on the tentacles:—

"From their extreme length and vast number they expose an extensive aggregate surface to the agency of the surrounding medium. They consist of hollow, flattened, tubular filaments, furnished with strong muscular parietes. Each of these hollow band-like tentacles may be rolled longitudinally into a cylindrical form, so as to enclose a semicircular space, if they only imperfectly meet. This inimitable mechanism enables each filament to take up and firmly grasp, at any point of its length, a molecule of sand, or, if placed in a linear series, a row of molecules. But so perfect is the disposition of the muscular fibres at the extreme end of each filament, that it is gifted with the twofold power of acting on the sucking and on the muscular principle. In addition to the two important uses already assigned to these tentacles, they constitute also the real agents of locomotion. They are first outstretched by the forcible ejection into them of the peritoneal fluid, they are then fixed like so many slender cables to a distant surface; and then, shortening in their lengths, they haul forward the helpless carcass of the worm." * The carcass of the worm is by no means so helpless as here described. It is true that the tentacles are employed to drag the animal along. You observe how that one is crawling up the sides of the glass, and now hangs suspended to the floating weed: but you may also observe him wriggling about his body with great activity, and by these contractions he is enabled to make progress, even when deprived of his tentacles.

There is a more serious objection, however, to be made to the passage I have just abridged. Dr

* WILLIAMS: *Report*, p. 194.

Williams—in common with most, if not all, anatomists—speaks of the *muscular* parietes of these tentacles. I venture to suggest that there is great inaccuracy in the term; and that the existence of these muscles is a pure assumption, assumed to explain the Contractility of these organs, in the same way as a nervous system is constantly assumed to explain some phenomena of Sensibility, although not a trace of a nerve can be detected by the highest powers of the microscope. The assumption is in each case perfectly needless, and very misleading. It is against all philosophy thus to assume the existence of a tissue no one can detect, to explain a phenomenon which may be otherwise explained. Nor is anything gained by declaring that the nervous tissue is in a "diffused state." This is making an assumption, and concealing it in a phrase. If I were to declare that gun cotton contained nitre, because gunpowder contains it; and if, when my statement was answered by repeated analyses proving no nitre to be there, I were to reply, "the nitre may not be detected by your analysis, because it is in a diffused state," you would shrug contempt at such chemistry. But this is precisely analogous to what is done daily with respect to nervous tissue. Men assume that all animals must have nerves; if the nerves are not visible, it is because they are "diffused." Now, this reasoning is not only vicious as logic, it is particularly vicious in Biology, where structure is of equal importance with composition. Nerve is a specific thing, having a specific composition, and a specific structure;

to talk of this thing as "diffused," is to talk of it as wanting one of its constituent characters; it is like talking of fluid crystals, or square circles. All this Dr Williams, I am sure, would be the first to admit, for he doubts the existence of nerves even in the Echinodermata; and I would ask him whether the tentacles of the *Terebella* are not assumed to have muscles, in accordance with the current notions that wherever there is Contractility the existence of muscles must be inferred? I put the question *as* a question merely. My own observations utterly failed to detect muscular fibres in the tentacles of the species I examined. On one occasion, indeed, they presented the aspect of circular fibres, which I thought must be the muscles Dr Williams refers to; but, on applying a power of 300 diameters, these fibres resolved themselves into simple corrugations of the investing membrane; and I can very confidently assert that in no single species of *Terebella* which has fallen in my way was there the slightest trace of muscle.

But enough of anatomy for this morning! The lovely lanes of Ilfracombe invite us, and we may cool our over-heated brows by a delicious breeze blowing over the Tors; or perhaps the noble sweep of Tenby sands seduces us to walk to Giltar Point. A bottle or two will be useful in either expedition; a small basket will be worth the trouble of carriage if we take the sands, for there was a gale last night, and who knows what may have been thrown up by it? And if you trust to your hands to carry all you may find, you will, perhaps, be the "observed of all observers," as I was, carrying a large Cuttle-fish in each hand, while some compassionate sailors superfluously assured me, "Them's not good to eat, sir!" Another day I transported a Dogfish through the streets—much to the horror of all the flounces, and the ineffable scorn of all the pink shirts and telescopes. You may be as indifferent to the stares and the scorns of flounces and telescopes as I was, but still I say, out of mere convenience, carry a phial, if not a basket. On one memorable afternoon we came upon, and almost stepped upon, an adder lying just outside the hedge. All is grist to the anatomist's mill, so I cut off the adder's head, and wrapped it in my pocket-handkerchief. Presently we came to a pleasant pond, the surface of which, with its varied *greenth* of scum, was so full of promise that there burst from me a sudden *Oh!* which startled, and not a little puzzled, a lazy countryman, taking his siesta by looking at nothing over a gate.

"Here's a pond!" I exclaimed, when reason got the better of emotion.

"Ah!" responded my companion, profoundly sympathetic.

The countryman was bewildered. Were we insane? or only Cockneys? There was a pond, sure enough, and as dirty a bit o' water as you'd wish to see; and what then? Were we frogs from the desert, that a pond should agitate us?

While he was cracking this very hard nut, harder

than his own Devonshire skull, I had emerged from the bitterness of self-reproach at having forgotten a phial, into the clearness of triumphant resource. Seizing a large dock-leaf and converting it into the rude resemblance of a bag, I hooked up with my stick a string of tempting scum, packed it up in the leaf, and walked away wealthy. To his dying day that countryman will recount, to all who will listen, the inconceivable fancy of the gentry folks, who carried off the filth of a pond in a dock-leaf. A queer start, warn't it?

Where shall we ramble? At Ilfracombe the question is really puzzling, because so many lovely walks solicit you. Go where you will, you cannot miss a lovely walk, that is some comfort; but there is an embarrassment of riches. Towards the close of spring, when the trees are in full leaf, but still keep their delicate varieties of colour—varieties lost in the fulness of summer, to be regained with even greater beauty in autumn,—at this time, when the furze is in all its golden glory, perpetually tempting one to pluck a tuft of blossoms as the largest specimen ever seen, and scenting the air all round, Ilfracombe is enchanting. So it is in summer; but the loss of the furze is almost like the fading away of the evening red. Contemporary with the furze is the lovely primrose, here seen to perfection, covering the hill-sides with pale stars, almost as plentifully as butter-cups and daisies elsewhere. In such a season, the walk to Lee seduces with its beauties of rocky coast and wooded inland hill; or the woods of Chambercombe lure you into their coolness. When the sun is broiling in cloudless blue, the coolness of a wood, in which the sunbeams only flicker through branches, and elicit all their beauties, forms a pleasant retreat; and before you reach Chambercombe the eye has been delighted with perpetual landscapes. There is a lane leading into a farmyard—a Devonshire lane, remember—which will long hold a place in my memory. Close to the gate of this farmyard there is a spring which is a perfect miniature of some Swiss "falls." It spreads itself like a crystal fan on successive ledges of the hedge-bank, until it reaches a much broader ledge, where it forms a little lake on a bed of brown pebbles; then down it goes again till it reaches the road, where it runs along a tiny, happy, babbling stream. One of the endless charms of these lanes—as of all mountainous districts—is the frequency of the springs, glossy with liverwort and feathery with fern, making a pleasant music day and night. Passing through the farmyard, where the pigs wallow, and grunt sensual satisfaction, and the cows look at you with bovine stupidity, you come upon a widening of the lane, where several gateways meet, and here the exquisite wild-flowers, everywhere so abundant, seem more than ever luxuriant. What a perfect bit of foreground is that! A few rough mossy trunks lying against the tufts of fern, and a quiet donkey stretched across the lane in "maiden meditation, fancy free;" it is one of those exquisite nothings which somehow affect you more than a fine landscape. At least it so affected us; and this was surpassed a little further on, when we came to a spot where a brook runs brawling across the lane, and a wooden bridge allows those to pass who prefer not wetting their feet. A rough hurdle is fixed up where the brook gushes from the field into the lane, over brown stones, which it polishes into agate. Against the little bridge rises a tree, and all round its roots by the brook-side are varied tufts of fern, and gems of wild-flowers. How I wished to be a painter that I might sketch such "bits" as these, and not let enthusiasm evaporate in ohs! From this brook a step or two brought us to a shabby house, bearing the reputation of being haunted, its broken windows rag-mended. I never saw the ghost; but I always saw a huge, divinely-awkward puppy, as happy and affectionate as puppies usually are. I could not get my companions to sympathise with me in my love for puppies in general, or in my wish to encourage the advances of this one in particular. *De gustibus.* There are people who don't like poetry; there are others indifferent to puppies. After a valedictory caress to this floppy acquaintance, we passed on into the woods, and while seated under delicious "umbrageosity," I soothed myself with a Latakia cigar, and contemplated a beautiful caterpillar spending its transitional life on a branch, happily knowing nothing of transitions. Pleasant was the murmurous sound of insects, pleasant the ripple of water, pleasant the glinting sunlight, and the broad reposing shade, but pleasant above all was the charm of interchanging thoughts.

Yes, Nature is very lovely, and speaks to us in soothing tones; but Human Nature has a holier accent still.

Another favourite walk was to Watermouth and Berryn Narbor, over the edges of majestic cliffs, revealing inlet after inlet, each differing in its wealth of colour, each a picture, till we passed into what are called the "meadows," really a noble park, through which runs a stream fringed with wild-flowers, and clear as crystal; every twenty or thirty yards the stream falls over an artificial precipice of stones, making a dulcet music. The slopes on each side are richly wooded; and the sequestered silence of this spot adds to its many charms. Who has not felt the deep peace which settles on the soul, when one is lying in the long grass beside a stream, under a summer sun, no sound of traffic, contention, or care to vex or sadden? Who has not sat upon a gate, less to rest than to enjoy the peaceful idleness of noon, and looked upon the marvellous forms of life active around him, dreaming all the while of pleasant scenes, which revisit the memory, or of pleasant hopes rising, "like exhalations of the dawn." In such a mood we one day rested on a gate under the trees beside this stream; presently a blind man felt his way also to the gate, and rested there. We spoke to him; he told us with that sluggish iteration characteristic of the countryman, that this was a fine healthy spot . . . yes, a very healthy spot . . . a healthy spot. And he held down his head; alas! it was useless for him to hold it erect, fronting the lovely scene. Saddened by his presence, we soon moved on, and returning over the cliffs, we came upon another human being, with eyes closed to the beauty around, but closed in sleep, not blindness. A little girl, not more than eight years old, was stretched along the path, her rosy cheek resting on her little arm, which rested on the bare rock. How fast she was! but, as Shakespeare says, "Weariness will snore upon a flint," and here was wearied innocence sleeping on a flint, the summer sun pouring down its rays upon her, and also on the milk, which stood in a can by her side, and which perhaps was not much benefited by this rest in the sun. All I know is, that the picture was very touching, and I placed a penny in the child's half-closed hand that she might find it on awaking. She would think some fairy placed it there.

On reaching home there was dinner, to which two words had to be said, not contemptuous, believe me; and then coffee and cigar, with the serenities thereon attending: and then a stroll among the vases, for the inspection of my pets; and a stroll in the garden where I could inspect the pans under the verandah; and then study; and then with limbs weary and eyes drooping, to bed:

"So runs the round of life from hour to hour."

CHAPTER III.

LIVE ANIMALS SENT BY POST—THE MORTE STONE—"WANTED, A TIN CANISTER!"—THE BORING MOLLUSCS, PHOLAS, TEREDO, AND SAXICAVA—HOW IS THE BORING EFFECTED?—APPLICATION OF EMBRYOLOGY TO PRACTICAL PURPOSES — RESPIRATION OF THE PHOLAS—DREDGING—THE SEA A PASSION—THE DEAD CEPHALOPOD AND ITS COLOUR-SPECS—NEED OF A DOCTRINE—MORMON PREACHER—LOGIC OF ZOOLOGISTS—CLEOPATRA'S PEARL—PRODUCE OF A GALE—THE EOLIS: ITS STRUCTURE; HOW DOES IT BREATHE? REASONS FOR REMOVING THE EOLIS FROM THE NUDIBRANCHIATES — WATER MINGLING WITH THE BLOOD.

Do you ever send live animals through the Post-office? The question may startle, perhaps, but the thing is often done. Only three days ago a brother naturalist sent me a couple of dozen Sea-Anemones, stowed among weed in a tin canister, which formerly contained a powder unblushingly sold behind a Christian counter as veritable coffee. The process is simple enough, when tin canisters are at hand, little as the excellent Rowland Hill contemplated such an adaptation of his postal reform to the exigencies of naturalists; but the process is less simple when you are temporarily abiding in a place so utterly provincial that little in the nature of tin boxes is to be had for money, and nothing at all for love. Such a place is Tenby.

But, first, let me tell you what made me desiderative of tin boxes, and indignant with Tenby for its want of resources. At Ilfracombe the orange-tentacled and orange-disked Anemones, by Mr Gosse christened *Actinia aurora* and *Actinia venusta*, are unknown, and, of course, prized all the more on that account. Is not everything valued for its rarity? There is, however, not many miles from Ilfracombe, a terrible reef running far out into the sea, bearing the sombre name of the Morte Stone, on which many a tall ship has been wrecked, and which, inaccessible from the land, is visited only by naturalists and gulls. We—I mean the naturalists, not the gulls—found Morte Stone well worth the visit; and while scrambling over its desolate ridges, the spray of a heavy sea dashing from either side in our faces, and a noonday sun pouring down its fierce passion upon our heads, as we clambered over rocks so black with mussels that you could not for yards have inserted a penknife between them, A., with his coat off, emerged from under a ledge, meeting B., no less jubilant, both holding up specimens of the orange-tentacled Anemone, hitherto supposed to belong exclusively to Tenby. This was the first time I saw the sunset-flame on the tentacles of this Anemone; and when at Tenby, remembering the delight with which B. carried home the novelty, it was natural that I should wish to send him a few of the beauties expanding their tentacles in my vases. But how? One cannot wrap a moist and mucose animal in note-paper, and expect it to reach its destination like an invitation for dinner, or the request for a "trifling loan;" and the damp sea-weed which will keep the animal alive, requires some covering to keep itself damp. I tried a card-board box, well padded with weed, wrapped it in paper, and committed it to the tenderness of a paternal Government and a reformed Post-office, with this warning inscribed in majestic calligraphy—

WITH CARE: LIVE ANIMALS!

I thought the Lacedemonian brevity and the note of admiration might have their effect. But, it is painful to confess, that Post-office clerks appear to be imperfectly versed in the rudiments of zoology; or perhaps they pay slight attention to the literature of Inscriptions; at any rate they stamped with a vigour which completely squashed the card-board box.

The next time, I determined to follow my friend's advice, and send the animals in a tin box, which, of course, seemed the easiest thing in the world, until the trial was made. I ransacked Tenby in vain. I asked everywhere—I asked at impossible shops—I even tried the bootmaker: he could not supply me. I offered money, and hinted love; but no tin box could Tenby produce. The article was mythical. Tenby had mustaches and parasols in prodigal abundance; pony-chaises and sailing-boats obtruded themselves at every corner; the streets were full of formidable young gentlemen from the fashionable parts (of Bristol), and nurserymaids with prize babies; these, and much more that was sublimely useless, Tenby had at your service;

but a tin box it knew not, except by vague report. Tenby has not even a banker; to get a cheque changed you must ride to Pembroke; why, then, expect it will have a tinman? Imagine my impatience, my disgust! I'm afraid I used strong language. At last a brilliant conception made my pathway clear. In a grocer's shop there were cases of ginger-nuts for sale; these cases were of tin; they were larger, much, than my requirements; but this was no occasion for drawing fine lines. The nuts were edible, the case transportable. An investment was straightway made, and my agitated mind was once more at peace.

The case was large enough to contain, besides a quantity of Anemones, a wide-mouthed bottle, in which I had consigned a fine specimen of that boring bivalve named *Pholas dactylus* (Plate II., fig. 3), three of which had been brought to me in a lump of wood, wherein they had bored themselves a local habitation. Although these Molluscs live in rocks and wood, they seem to flourish perfectly when removed, and left in sea water. I risked one in the experiment, and was uneasy, next morning, at finding he had elongated himself to more than half again his original size; but observing the currents were still active from his siphon, and that, on being touched, he shrunk to his original size with great sensitiveness, I concluded he was healthy,— a conclusion supported by observing precisely the same phenomena exhibited by the two Pholades still in the wood. As, therefore, the Pholas lived out of his woody home, and as I had three specimens, I could do no less than send one of them to an amateur; and thus it came that I despatched the wide-mouthed bottle in the tin case, which arrived without accident. After keeping my two borers for some time, one of them fell a victim to anatomical curiosity; as for the third, *l'appétit vient en mangeant,* and I dissected him also.

The reader has doubtless heard about the boring Molluscs, of which there are several different kinds, all curious to the philosopher, but none very interesting to keep. One species, the *Teredo navalis* (Plate II., fig. 5), is a formidable fellow, unloved of ship-owners, since many a ship has been known to split in the open sea, no one on board having suspected that the planks had been thoroughly drilled through and through by this patient borer. The hardest oak, nay, even teak and sissoo woods, are no obstacles to this mollusc. The chemical process which protects timber against decay is no protection against the Teredo. The animal always tunnels in the direction of the grain of the wood, and if in its course it meets with another gentleman engaged in the same process, it alters the direction of its course, so that a piece of wood attacked by many Teredos becomes completely honey-combed. In dockyards the defence has been to cover the woodwork under water with iron nails; and you may imagine how necessary some protection is, since not only have docks been perforated, but many years ago Holland was thrown into terrible alarm by discovering that the piles of her embankments had been riddled by these silent molluscs, and all Europe thought that the United Provinces were doomed. Other borers choose rock for their operations; and many a solid-seeming mass is so perforated by them, that the dashing of a stormy sea may scatter it in fragments along the coast. The fact of boring is familiar enough to every one who has noticed the Red-nose (*Saxicava rugosa*) peering from a thousand holes in the hard limestone, and squirting water as it retires on the first application of the hammer; but while the fact is undisputed, the source of the animal's power is still an unsettled question.

How these bivalves bore their way has been a mystery, mainly because zoologists have allowed themselves to be thrown off their balance by contemplating the stupendous results produced by creatures so insignificant. But after learning the history of the formation of coral reefs and islands, we begin to appreciate the influence of minute agencies continued through long spaces of time. The Teredo and the Pholas have no powerful organs, but they have patience; and as far as I understand the matter, it is clear that the disputes on this subject have been perplexed by the desire to bring forward some organ so powerful as at once to explain the animal's success in boring. Thus the latest writer, M. Aucapitaine,[*] imagines he explains the phenomenon, by bringing forward the hypothesis of an acid secreted by the animal, which corrodes the rock, or wood, and which is then rasped away by a slow rotatory motion of the shell. The

[*] *Annales des Sciences Naturelles*, 4ᵐᵉ série, 1854; ii. 367.

boring is thus supposed to be a combination of chemical and mechanical agencies. There is, however, one little difficulty in accepting this explanation, which the author has overlooked, as speculators are wont to overlook fatal objections: the *existence* of this acid has yet to be proved; its presence is indispensable to the theory, but, unfortunately, the fact of its presence is hypothetical. And when we have got tangible hold of the acid, we must prove that, while it has the property of attacking wood and rock, it has *not* the property of attacking the calcareous shell of the animal.

Very different, and far more philosophical, is the explanation of Professor Owen, whose opinion on all points carries authority. He shows* that the combined action of the muscular disc and the valves of the shell will produce the phenomenon. It may be paradoxical, and you will probably shake a dubious head on seeing the cavities bored by these Molluscs, and on being told that the soft muscular disc of the animal perforated them; you have no conception that, by licking limestone with never so much energy, you could wear it away; and yet, as Owen quietly remarks, "it is certain that the perpetual renewal of a softer substance will render it capable of wearing away a harder one, subject to the friction of such softer surface, and not like it susceptible of being repaired." Yes, here lies the whole mystery: the soft muscular disc is perpetually renewed, and the hard limestone

---

* *Lectures on Comparative Anatomy,* i. 520.

has no self-renovating power ; and thus, just as falling water wears away granite, by the incessant repetition of gentle blows, so do these Molluscs excavate rock, or wood, by the incessant repetition of muscular friction.*

Some practical man, who does me the honour to relax his serious mind over these pages, here declares himself supremely indifferent to this anatomical discussion. What does it signify to him *how* the Teredo bores into the wood ? He is none the better for that. It is enough for him that the nasty beast *does* it, and unless he can be told how it is to be prevented, he wants to hear nothing about the matter. As a practical man, he wants practical applications ; as for " theories," he doesn't care a silver fourpence for them. I will not turn round and humiliate him, by proving that of all blind theorists none are blinder than the " practical men ;" but will rather captivate his confidence, by showing him how the result he so earnestly desires is only to be obtained after a remote excursion into the obscure regions of science. *He* need not make the excursion, but he must wait till it be made ; for it is amusing to think that even so simple a matter as the destruction of these vermin defied all ingenuity, until Embryology came to our aid. I carefully abstain from mentioning that unusual term in his presence, but address a question to him :—

" Are these animals of separate sexes ?"

---

* VICTOR CARUS—*Jahresbericht über die im Gebiete d. Zootomie erschienenen Arbeiten*, i. 110—gives a dozen references to papers on this question.

He looks rather huffed, as he replies: "How should I know? and what does it matter?"

"It matters everything. And for your satisfaction I can tell you that they *are* of separate sexes."

"Humph!"

"A French naturalist, Quatrefages by name, has found that at certain seasons the female carries the eggs in the folds of her respiratory organs."

"The deuce she does!"

"And there they remain till the milt of the male, floating in the water, is washed over them, and fecundates them. Now this same Naturalist has brought his knowledge to bear upon the very question you are so interested in. He finds that a weak solution of mercury thrown into the water destroys this milt, and consequently *prevents* the fecundation of the eggs, nipping the young molluscs in the bud. By thus becoming a zoological Herod, and destroying the innocents wholesale, in a few seasons you may clear the docks of every individual Teredo, and your ships will be safe."

I see by the intelligent twinkle of his eye that he has seized this notion with decisive approbation, and from this moment begins to think there may be some use, after all, in Natural History. I almost feel tempted to show him my Pholas, although it is not a very interesting animal to watch. There is a somnolent lethargy, an otiosity of do-nothingism in its demeanour, surprising in one who bores through rocks as we tunnel the Jura. He will not even bore now. I have tried him in vain. He lolls his great length at the bottom of the pan, and declines the lump of wood placed before him. In fact, he does nothing but suck in the water at one tube of his siphon, and squirt it out of the other. Do observe that siphon or double tube, like a double-barrelled gun, the lining membrane of which is covered with vibratile cilia. The incessant action of these cilia draws the water in at the orifice of the upper and larger tube, along which it passes and reaches the gills, where the blood is aerated ; and then the water makes its exit from the under tube, in a steady current, visible to the naked eye. How this is performed is to me a mystery, for my dissections wholly failed in tracing *any* direct communication between the two tubes ; but that there must be some indirect communication is certain, since the evidence of the two currents, one of entrance, and one of exit, is unequivocal.* Look at the orifice of the upper tube, what a beautiful arborescent fringe encircles it !

* This disputed question has been finally settled by the investigations of Messrs Alder and Hancock, who find the communication takes place through minute apertures between the meshes of the gills themselves. "Each of the gill-plates consists of two laminæ united at the ventral margin, and likewise attached to each other in transverse lines running across the gills throughout their whole extent, and forming, in the interspaces, a series of parallel tubes which open into the dorsal chamber, and are thus in communication with the excurrent siphon. The minute reticulated blood-vessels of the branchial laminæ forming the walls of these tubes, are found, when examined by a high power of the microscope, to be open between the meshes, which are minutely ciliated, allowing the passage of the water into the tubes, and from thence into the anal chamber."—*Report of British Association*, 1851 ; Sections, p. 75.

You may see a picture of it in Mr Gosse's *Rambles on the Devonshire Coast* very well delineated. "The tentacular filaments," he says, "are numerous, each forming a little tree with pinnate branches, bearing no small resemblance to the flower of feathery branchiæ that expands round the mouth of a Holothuria. These branched tentacles are ordinarily bent down across the mouth of the tube, the longest of them just meeting in the centre; alternating with these are placed others of similar structure but inferior size; and the interspaces are occupied by others smaller still, and simply pinnate; so that when the whole occupy their ordinary transverse position, the smaller ones fill up the angles of the larger, and the branches of all form a network of exquisite tracery, spread across the orifice, through the interstices or meshes of which the current of entering water freely percolates, while they exclude all except the most minute floating atoms of extraneous matter."

The boatman has just called to say the boat is ready, and the Dredge at our service. In the previous chapters I described our hunting on the rocks, and picking up what gales might have thrown upon the shore; and the amateur generally contents himself with these resources, unless his desires, enlarging with his knowledge, urge him, as they did me, to follow more ambitious naturalists, and try dredging. He knows that in depths never laid bare by retiring tides there are animals of price. He knows that the oyster-beds are hunting-grounds where a single venture will bring him more than a month can properly examine. It is true that he may also know that he will be sick ; but, as Schiller says,—

> "Es wächst der Mensch mit seinen grössern Zweeken ;"
> (Our stature heightens with our heightening aims),—

and the hope of Molluscs makes man's stomach equal to the occasion. Our boatmen told us of one well-known Anatomist, who went out every day during his stay at Tenby, dredging as if dredging were his daily bread, always sick, no matter how calm the sea ; always suffering, but never daunted by wind or storm. Very amusing it was to notice the puzzlement of these honest boatmen at what they evidently considered a sort of inexplicable eccentricity in our thus throwing away our days, our money—and our breakfasts—in the pursuit of worms, oyster-shells, and weeds. Had we gone fishing, they could have entered into our hopes and enthusiasm ; had we sought for pearls in the oyster-shells, their sympathy would have been ready ; but that any sane man should be anxious for the rubbish which they nightly threw away when their nets were hauled in, and this not to eat the worms, not to sell them, but to put them in vases, and finally cut them open, *that* was inexplicable.

As we sailed through the sparkling waters, wafted by a pleasant breeze, we talked to the men, and tried to make them understand the kind of things which they might always bring us, and be certain of purchasers ; but although willing to oblige, and not at all indisposed to accept silver for a little trouble, although one of the men had once picked up a bunch of sea-grapes (eggs of the cuttle-fish), for which a gentleman gave him half-a-crown (a mad gentleman, clearly), these stolid fellows always fell back upon their ignorance. "Ah! if we only knowed the things." In fact, no bribe will move them. They cannot realise to themselves the conception that what they have for years thrown away as rubbish can possibly contain anything worth money. I repeatedly urged on them these simple instructions: Bring me *anything* alive (except fish) that you find in the net, and the chances are that I shall buy it. One Calamary (*Loligo*) was all that came.

We have reached the oyster-beds, and the Dredge is dropped into the water, plunging some fifteen or twenty fathoms, like a diver knowing what is required of him. On we sail, the line running out, the dredge raking the oyster-beds, and considerably retarding our progress in spite of a stiff breeze. At length it is time to haul in, and the men pull strenuously, till the Dredge appears, and a portentous mass of oyster-shells, dirt, stones, and Sea-Urchins is emptied in the bottom of the boat. We pounce on it, while the Dredge is once more cast into the water.

Up to this moment we have been superb seamen; Britons are the boys for ruling the waves! The colour has not changed on the cheek of the lady naturalist, who, astonished at her own fortitude (of stomach), declares dredging preferable to hunting on the rocks. But suddenly a change comes. This stooping to examine our treasures disturbs our pleasant serenity, and the well-known head-swimming and nausea which ensue, categorically tell us that, let Schiller declaim as he will, poetry won't control the liver; and however successfully Britons may rule the waves, the waves are extremely disrespectful to their rulers. *Que voulez-vous ?* The brain may be confident, but the liver is upset; heroism is futile against a chopping sea. We can't pretend to be superior to these weaknesses, and so we resign ourselves to sitting quiet, while the boatmen turn over the mass, and hand us objects for inspection, upon which we decide whether they shall be placed in the bucket of sea water, or returned to the oyster-beds.

Very droll it was, even in my languid state, to observe Jack stooping and fumbling among the oyster-shells, not knowing what his insane party might possibly think worth carrying home, and for his part thinking the whole as big a heap of rubbish as ever he saw.

"This here any use, sir?" he inquires, handing me a huge oyster with an unexpressed sarcasm in the question. I turn my green countenance towards it, and suddenly forget the qualms, exclaiming—

"Use, my dear fellow! of course it is. Why, it is a perfect colony of animals."

"Well, sir, you knows."

And he drops it into the bucket, plunging his hands once more among the mass. That oyster, besides the Polypes and Sponges growing on it, bore at least a dozen Terebellæ, an Ascidian of exquisite colour, innumerable Serpulæ, and a beautiful Sabella.

"Stop! what is that you're going to throw away?"

"Only a bit of dirt, sir."

"Let me see it. I have known bits of dirt turn out to be curious animals."

Jack, now fairly bewildered, and expecting probably that the next thing he will be asked to hand me will be a bubble of foam, stretches out his honest fist, and places on the seat a small lump of sand, having no definite shape, and looking no more like an animated creature than the mud-pie which ingenious youth delights to construct. I know it at a glance to be an Ascidian (*Molgula arenosa*), for only last week, while scrambling over the rocks, I looked into a shallow pool, on the sandy bottom of which there was one of these sand-lumps alone in its glory. I cannot tell what made me suspect it to be an animal. The mind sees what the eye cannot. Do we not distinguish a friend by a certain undefinable something long before he is near enough for us to distinguish his dress or his features? With the same mental perception one learns to distinguish an animal, even when one has never seen it before. I had never seen or read of this Ascidian. I did not know it to be an Ascidian; but, detaching it from the rock, I popped it in my bottle, convinced that it was an animal of *some* kind; and on coming home, I began to scrape away the sand till I came down to a membrane. I then cut the mass open, and found an Ascidian, which had so completely coated itself with sand, that the sand became part of its integument.

Having turned over the heap, and shown me one by one every shell or weed, Jack now began to clear the boat, and to haul in the Dredge once more. After a few hauls, our bucket was sufficiently stocked, and we sailed homewards, skimming the surface with a net in the hope of capturing some jellyfish, but none appeared. We ought to have been in high spirits; but whatever consolation may have been in the thought of the bucket, we were not hilarious, I pledge you my word, as we scudded along, green and silent. We reached the pier-steps at last. Jack carried the bucket after us, and the perplexed vassal at the lodging-house brought down a foot-pan, into which the contents of the bucket were emptied. I should not like to inquire too closely into that vassal's private opinion of me and my pursuits. The next day, when I met Jack, he was gratified to learn that the result of our dredging had been highly satisfactory, as indeed it was; for besides abundance of known animals, I had found two entirely new genera of Annelids.

Sea-sickness is not an agreeable sensation. While enduring it, we all vow never again to brave it without urgent necessity; and yet the next day we forget our resolution, and step lightly on board, mentally singing—

"The sea! the sea! the open sea!
I am where I would ever be!"

And if we do this merely for the sake of running to

France or Germany, where we have been before, it may be expected that the more powerful attraction of marine treasures will entirely conquer hesitation.

The fact is, the sea is a passion. Its fascination, like all true fascination, makes us reckless of consequences. The sea is like a woman: she lures us, and we run madly after her; she ill-uses us, and we adore her; beautiful, capricious, tender, and terrible! There is no satiety in this love; there never is satiety in true affection. The sea is the first thing which meets my eyes in the morning, placidly sunning herself under my window; her many voices beckoning me, her gently-heaving breast alluring me, her face beaming with unutterable delight. All through the day I wanton with her; and the last thing at night, I see the long shimmering track of light from the distant beacon thrown across her tranquil surface—dark now, and solemn, made more desolate by the dark and silent hulls of anchored vessels, but beautiful even in her sombre and forlorn condition. I hear her mighty sighs answering the wailing night-winds. She lures me to her. I cannot go to bed, but wander along the sands, and gaze upon the solemn gloom, stretching mysteriously afar. I walk down to the quay; all is silent, except in one boat, where a knot of men are just about to start for their night's fishing. They will be out all night, toiling through the terrible waters to gain a few shillings. I bid Jack bring me a cuttle-fish if he can. "Good night, sir."—" Good night, and good cheer." And away the boat speeds into the night. It is soon invisible; the plash of the oars ceases to reach my ears. There is something pathetic in the thought of these men nightly braving what they brave, and totally insensible to the poetry of their situation, which might make it something better than a mere venture for a few pence. My thoughts are sad to-night. I wander on, and the waves come to greet me, but the image of that boat disappearing through the darkness will not leave me. Life seems so sad, so transitory, so ineffectual, and Nature so pitiless and calm.

The next morning all such thoughts have vanished like uneasy dreams. Nature is joyous, clear, sunny; my mistress yonder is sparkling and singing in the light; white sails dot the distance; the busy hum of men rises on every side. I go out on the sands, and at my feet the tide throws a Calamary (*Loligo*) with which I rush back to my lodgings in great glee. A pie-dish of sea water receives the welcome Cephalopod; but he is dead, and will show none of his ways. Yet what is this? the colour-specs are coming out on the skin, like stars appearing at night; and now the whole surface, which was pearly-white, is of a variegated hue.

I had heard of this fact before, but actual observation gives one very different feelings from those of mere acquiescence in a fact. The colour-specs continued to come and go, much to my puzzlement; nor could I gain much light from any books at hand. M. Alcide d'Orbigny, who has studied Molluscs for twenty years, especially the Cephalopoda, ought, of course, to be consulted, and you shall hear what he says: "The colour is as unchangeable as the diverse impressions of the animal. It consists of a complicated system of globules of various colours, red, brown, and yellow, placed under the first layer of the epidermis. These globules have each a pupil which contracts and dilates, forming now a large irregular spot, and now diminishing to a mere point. It is easy to understand, therefore, how the animal which, when the globules are dilated, is of a dark colour, becomes almost white when these globules are contracted. The contractability of these globules *therefore always depends* on the will of the animal; and he varies his colour from brown to white with remarkable vivacity according to his own will." *

Attention is called to this passage, as one among the numerous illustrations of that serious want of a doctrine to guide investigations, which is the greatest obstacle Zoology has now to contend with. The mass of facts which has been accumulated is of astonishing extent; but the philosophy which should be evolved from them, which should co-ordinate them, and serve as a torch to guide zoologists in all inquiries, is still in the most meagre condition. To confine ourselves, for the present, to the case before us, is it not remarkable that a man so eminent as M. d'Orbigny should have written, and some other men acquiesced in, a passage so preposterous as the one just cited? Where was the biological philosophy which could conceive the contractility of pigment globules as dependent on the "will of

* D'Orbigny: *Mollusques vivants et fossiles*, p. 113.

the animal;" especially when such a leap in logic had to be taken as is taken at the "therefore?" Physiologists are, indeed, extremely facile in their admission of "the will" to explain what they do not understand; but we must marvel that direct observation did not utterly discountenance its introduction here; for it is quite certain that hundreds must have *seen*, if they did not *observe*, what most attracted me in the matter— namely, the appearance and disappearance of the colour-specs in the skin of the *dead* animal; and even the most metaphysical of zoologists would hardly attribute volition to a corpse. The observation of this one fact might have led to further investigation. On placing a small strip of the skin under the Microscope, I was surprised to find two or three of these colour-specs expanding and contracting with great vigour. At first I thought it must be an optical illusion, but on close attention it became too decided for doubt; and not suspecting the truth, I concluded that some animalculæ were imbedded in the tissue, and that their movements produced this apparent activity of the globules. To settle these doubts, two other strips underwent examination; in both of these, all, or almost all, the specs were in activity, shooting out prolongations, and retracting again—two specs sometimes seeming to run into one, but really overlapping each other, and sometimes a point not bigger than a millet seed expanding to the size of a sixpence, growing fainter in colour as it expanded. This was decisive. If the globules in a strip of skin taken from the dead animal manifest precisely the same contractions and expansions which they manifest on the living animal, it is clear that their activity does not depend upon the "will,"—a conclusion which elementary principles of Biology ought to have made self-evident *a priori*.*

But it is not in Zoology only that logic is courageously assaulted by man's "large discourse of reason." If "reasoning correctly on false premises" be the true definition of madness, we are all more or less madmen; although we are all "astonished" at the insanity which we do not share. Last evening this was brought before me in half-sad, half-ludicrous aspect. We were smoking, in the indolent beatitude of digestion, when

---

* DELLE CHIAJE, in his *Descrizione e notomia degli animali invertebrati della Sicilia Citeriore*, i. p. 15, says that the expansion of these colour-specs is due to an expansile liquid, allied to blood—*espansile umore (ematosina ?)* which is contained in the vesicles, and which is probably in relation with the blood-vessels and the *rete Malpighi;* and he suggests that its contractions and expansions may depend on respiration. But the fact, recorded in the text, of a strip of skin taken from the dead animal exhibiting the same contractions and expansions as those exhibited by the skin of the living animal, shows that the contractility of these vesicles is independent of any such cause. Kölliker has shown that the contractions are produced by pale muscular fibres. Some doubt, however, is permissible respecting their *muscular* nature. As Charles Robin says, they are perfectly homogeneous and extremely fine; moreover, they are not capable of being isolated *as* fibres, "en sorte qu'il serait plus juste de dire que les chromatophores (coloured specs) sont entourés d'une substance homogène contractile, fibroïde; les fibres non isolables, et dont par suite le diamètre exact ne peut être donné, sont disposées par faisceaux." † No one has explained how these retain their contractility so long after all the other muscles.

† Note communicated by LEBERT in his "Mémoire sur la Formation des Muscles."—*Annales des Sciences*, 1850, p. 172.

a choral howl disturbed the quiet of the evening air. P., lolling over the balcony, and allowing the "demure travel of his regard" to sweep the horizon in search of the yacht which was to fetch him away, informed us that the howling came from three itinerant preachers about to edify a group of fishermen on the quay. I begged him to shut the window; this being my protection against the outrage of a German Band, which daily for six weeks had played " Partant pour la Syrie," " The low-backed car," "The Red and the Blue," and " God save the Queen"—never anything else, and always pitilessly out of tune. But P.'s sense of the ludicrous overcame his musical susceptibility, and condemned us to hear the hymn. Shortly afterwards, the preaching began, and as we ascertained that it was Mormonism then being expounded, we resolved to go out and be edified. It was worth the effort. Standing on a chair was a young man, scarcely above twenty, swinging his arms about, and flinging forth in harsh ejaculations a torrent of repetitions and abstractions, quite distressing to listen to, from the total want of anything that could arrest the interest of his audience. Open-air preaching is meant to coerce the attention of those who will not go to church; but this Mormon preacher never once alighted upon a phrase which could awaken an idea in the minds of those he addressed; so that we marvelled why he should have been chosen as a preacher of a doctrine which addresses the worldly interests. On each side of him stood an Elder—and I wish I could paint the portraits of these Caryatides of imbecility. One was a well-washed middle-aged man, who may have been a sentimental tailor; he rested his elbow on the chair which served as a pulpit, and, inclining his head, allowed his finger to indent his cheek. The other was a short, tawny, grey-haired man, who must have been a cobbler troubled with metaphysical misgivings. It is to be presumed that they were edified by the preacher's rhapsodies and repetitions; the audience was utterly unimpressed. Indeed there was what P. called a troublesome foreground of boys and girls, fighting, laughing, jeering, beating tin kettles, and otherwise exhibiting the moderate sensibility of their *fibre religieux;* but the background of men and women (of course with babies) was more orderly. They listened in respectful silence, but with no appearance of sympathy. A grey-haired fisherman standing beside me said to a woman at his left: "He doesn't speak according to Scripter. Some things is according to Scripter; but some is not." He spoke in a quiet assured tone of authority, and his was the only criticism I heard.

This is a digression, and has only a remote connection with the imperfect logic of zoologists, a subject on which, if I had greater authority, I would discourse at length. Not that I suppose zoologists to be less logical than other men; but simply that the Science of Life being so much more difficult than any of the Physical Sciences, it is more in need of a rigorous code of principles. The Astronomer, the Physicist, and the Chemist, are subject to restraints which the Biologist seldom condescends to regard. No speculative Chemist is allowed to call a substance an acid which will unite with no base, which exhibits none of the properties of an acid; no Physicist is allowed to assume the existence of electricity, where none of the conditions of electricity exist, and none of the phenomena (except those to be explained) are manifest. But we who study Biology in any department, whether Physiology, Zoology, or Botany, are allowed by the laxity of current practice, and the want of a doctrine, to call a coloured spec an eye, in the absence of all proof of its having the structure or functions of an eye; we are allowed to assume the existence of nerves, where no trace of a nerve is discernible; we are allowed to drag in "electricity," or the "will," as efficient causes of anything we do not understand; and we fill Text-books and Treatises with errors which give way before the first sceptic who investigates them.

We are very sheep in our gregariousness in error. When one bold or stupid mutton takes a leap, all leap after him. It is rare to find men doubting facts, still rarer to find them doubting whether the facts be correctly co-ordinated. Our books are crowded with unexamined statements, which we never think of examining. Do we not all believe that the magnificent Cleopatra, regardless of expense, dissolved in her wine-cup a pearl of great price, as if it had been a lump of sugar? Is not the "fact" familiar to every one? Yet, if you test it, you will find the fact to be that pearls are *not* soluble in wine; the most powerful vinegar attacks them but very slowly, and never entirely dissolves them; for the organic matter remains behind, in the shape of a spongy mass larger than the original pearl.

"Forewarned, forearmed." Students once having their attention called to the necessity of scepticism in Zoology, will soon find abundant occasion for its exercise. We should as much as possible keep the mind in a state of loose moorings, not firmly anchoring on any ground, unless our charts are full of explicit detail; not *believing* (but simply acquiescing, and that in a provisional way) in any fact which is not clear in the light of its own evidence, or which, in default of our having verified it for ourselves, has the trustworthy verification of another. This sounds like a truism, but it is not my fault if it be necessary to enforce a truism. The adoption of such a rigorous scepticism would revolutionise Zoology. It will not be adopted by the majority, because it will give great trouble, and men dislike trouble; but the more restless and rebellious spirits, of which there is always a proportion in every sphere, will "scorn delights and live laborious days," in subjecting accepted statements to rigorous verification.

I came down to the coast as an amateur, ignorant, but anxious to learn, and not simply to seem to learn. For this purpose it was not enough that I should know what was said respecting the simpler organisms, but it was necessary, as far as possible, to understand the grounds on which each statement was made. Many and many an hour was spent, but not uselessly, in verifying what every tyro knew to be the fact; but also many an hour was spent in making clear to myself, not only that certain accepted statements were errors, but how they became accepted. As an example may be mentioned the respiration of sea-slugs (*Eolids*), in the investigation of which I was favoured by fortune as not many are favoured, namely, by having abundance of material to work on.

There had been a heavy gale all night, and the wind was still high; down we went to the Tunnel Rocks to watch the tide come in. It was a glorious sight to contemplate the impetuous sea plunging upon the shore with ever-accelerated velocity, rising in wrath, and leaping over the reefs with mighty bounds, roaring, hissing, groaning, sighing. We stood with our backs leaning against a wall of rock, the spray leaping up into our faces. At length a black mass appeared upon the swelling height, to be lost again in the ridge of foam, and then to reappear; onwards it came, struggling with the waves which tossed it and tumbled it to and fro, till we descried it to be a bit of wreck. In one instant the sense of the picturesque was submerged by a rush of zoological expectation. We sprang down on the shore, anxiously awaiting till the prize should be flung at our feet. It turned out to be the bit of an old cask, which must have been long under water, for it was as black as ebony, and literally covered with Polypes and eggs of Eolids. You may be sure it was welcomed with jubilant shouts, as were the masses of weed thrown up at the same time, also covered with Polypes and ova. We departed with the feelings of men who have just heard of a legacy. Next morning I found the treasures greater than our expectation; not only were there thousands of ova, but scores of delicate and tiny Eolids of different species were found floating in the water, or crawling among the Polypes. It took hours to remove these delicate creatures into separate vases and bottles, and then to contemplate them with hungry enthusiasm—which you will appreciate if you have been ever fascinated by the study of Development, and suddenly seen abundant material within reach ; or, if you have been anxious to solve some problem which only abundant observation could help you to solve. I was troubled about the respiration of the Eolis, not feeling at all satisfied with what is taught in the schools ; and here were more Eolids thrown into my hands than most men ever see in a lifetime.

The Eolis (Plate II., fig. 1) is a sea-slug, but in spite of this ill-sounding name, the sea has few creatures more elegant in form, more exquisite in colour. In size it ranges from one-tenth of an inch to three inches in length. The commonest and least handsome species, *E. papillosa*, is about an inch or an inch and a half long, the back being densely studded with slate-coloured club-like projections, called branchial papillæ, so like in colour to the rock it crawls over, that often only an experienced eye can detect it. The more elegant species, such as *Eolis pellucida* or *elegans*, or

Lansbergii, should be sought for every Aquarium, care being taken to keep them out of the way of the Anemones, which they mercilessly attack. Believe in no woodcut representations of these exquisite creatures; all woodcuts are libels. The plates of Alder and Hancock's magnificent monograph approach as near to the beauty of nature as can be expected of plates; but even they necessarily fall short of the delicacy of tissue and witchery of colour often displayed by these animals. There is nothing but actual possession which ought to satisfy you; and possession is not difficult.

We delicately lay open the back to expose the stomach, which lies on the dorsal, not the ventral surface. We shall find the Eolis rejoicing in a digestive apparatus as perfect as that of an Alderman, but somewhat different in structure. Stomach, properly so called, it can hardly venture to claim; for that pyriform pouch which you observe, is rather an expansion of the intestinal tube than a distinct organ. Observe how this tube is continued along the whole length of the body: in some species it is wide and tapering; in the one before us it is more constricted; and be particular in noticing how this tube gives off pairs of branches, which again subdivide into smaller branches, and run up into those club-like projections, called dorsal or branchial papillæ, the cavities of which they almost entirely fill, sometimes as mere dilations, sometimes with shrub-like arborescence. Having thus traced the stomach, the intestine and its ramifications, we must now look out for the next important organ of the digestive apparatus, the Liver. It is useless seeking for it; the Eolis has not reached that stage of animal development which imperiously demands a special biliary organ; it can transact all digestive requirements by the aid of biliary vesicles scattered along the lining membrane of those intestinal branches which we saw filling the cavities of the papillæ. In a word, we here meet with the rudimentary and initial condition of the liver, nothing more than a few hepatic cells.

You understand, therefore, that these papillæ covering the back, and bristling up like quills upon the fretful porcupine when the Eolis is irritated, are hollow tubes, containing prolongations of the intestine, and biliary cells. By many anatomists it is thought that these tubes are biliary ducts, and perhaps in some species this may be so, but that it is not so in all, I have positive evidence. I was one day examining some Polypes, when my attention was diverted by something granular, contractile, indistinguishable in shape. On extricating it from the branches of the Polype, I found it to be a white oblong jelly-like creature, about the size of a pin's head. On placing it under the microscope, I saw that it was an Eolis, but whether of a yet undescribed species, or only the young of some known species, imperfect knowledge did not enable me to decide. The extreme transparency of its tissues was such that I could observe it with satisfactory precision. It had only eight papillæ on either side, but these were very large, each of the central ones measuring at least a sixth of the whole animal. This was another fortunate circumstance, enabling me to detect the passage of the dark granules which almost filled the cavities, to and fro from the intestine, with which these cavities were in direct communication. All doubt was impossible: there was the food oscillating from the intestine into the papillæ, and back from the papillæ into the intestine; and this oscillation, I observed, did in nowise depend on the contractions and expansions of the papillæ, but wholly on the contractions and expansions of the body; for sometimes the granules ran up into the cavities when the papillæ expanded, but sometimes they remained in their previous position.

You see that the papillæ are gastro-hepatic organs, or, to speak less technically, that they are parts of the intestinal and biliary apparatus; but you see nothing to warrant the accepted notion that they are respiratory organs. I certainly saw nothing of the kind. It was a doubt which early forced itself upon me. Zoologists class the Eolis among Nudibranchiates, but I could detect nothing like a gill in these said naked-gilled molluscs; and a series of dissections served to transform doubt into a conviction, and satisfactorily proved the term "branchial papillæ" to be altogether erroneous. These papillæ have neither the specific structure, nor the anatomical connection of gills. Various as gills are, they have uniformly a system of vessels carrying the blood to them for aeration, and a system of vessels carrying the aerated blood from them. Have the papillæ of the Eolis such vessels? Nothing of the kind. The blood they receive, which is not more than other parts, is received into sinuses, as it is in the cavities of the body; and it is not more aerated there than it is in every other part of the mantle from which the papillæ rise. If, therefore, we can detect in these papillæ neither the ordinary structure of gills, nor the vessels which carry blood to and from gills, it is eminently unphilosophic to call them gills, and to class the Eolids among Nudibranchiates.

Do but examine one of the other Nudibranchiates—say a Doris—and you will there find the very characteristics wanting in the Eolis. It has a gill, distinct, unmistakable; although even here the gill performs but a small part in the aeration of the whole blood. According to Alder and Hancock, only that portion of the blood which supplies the liver-mass goes to the gill; but small as the part may be, the organ is distinctly recognisable, and to compare it with the dorsal papillæ of the Eolis is to demonstrate that two such dissimilar organs cannot play the same part. Indeed, the Doris seems to me higher in the scale of organisation than the Eolis, although less active in its movements. It has a specialised liver, a more perfect vascular system, the commencement of a respiratory system; and it has *not* the arborescent intestines which the Eolis has in common with the *Planariæ* and *Pycnogonidæ*. I should propose, therefore, to remove the Eolis from the Nudibranchiates, and call it Abranchiate.

How, then, does the Eolis breathe? He does not

breathe at all. But lest this paradox should disturb you too much, I will soften the blow by adding, that when we talk of an animal breathing, we mean, or ought to mean, that it employs an organ, or group of organs, for the aeration of its blood; and when the animal is of so simple an organisation that it possesses no such organ, although the aeration takes place quite as well, it does so in a different manner. Respiration as an *animal function*, and Respiration as a general *property of tissue*, are incessantly confounded in our loose language; but the distinction should always be borne in mind. The ultimate fact of Respiration is the interchange of gases, and this may be effected in many ways; but although the final result is similar, there is great difference between the property which all living tissues, animal and vegetable, have of exchanging carbonic acid for oxygen, and the function of the special apparatus by which the exchange is brought about; just as there is a wide distinction between the general property of *Assimilation*, and the special function of *Digestion*. The Eolis we are considering must have its blood aerated; but the means by which it is aerated do not come under the term "breathing." In many of the lower animals aeration is performed entirely by the surface, the air or water directly bathing the delicate tissues, and bringing to them the necessary supply of oxygen, without the intervention of any special apparatus; just as food is brought to their tissues without the preparatory labour of arduous digestion. The Eolis has not only a delicate surface, covered with vibratile cilia, which permits the aeration of all the blood circulating at the surface, but it has also a system of water-carrying vessels round the margin of its broad foot, or fleshy disc, through which water is carried into the cavities of the body, and there aerates the blood,—at least I assume this latter part of the description to be true of all Eolids, as it is of the Doris, although I only observed the vessels in one species, not having thought of seeking them until my last animal was under the scalpel.

Having made thus much clear to myself, I found that I was not so heretical as I fancied, but that, for the main facts on which my argument rested, the authority of even Alder and Hancock could be claimed. These admirable authors distinctly say, " the respiratory function appears to be partially specialised in the dorsal papillæ, which, usually exposing a large surface, are covered with strong and vigorous vibratile cilia. [Not so strong as the cilia of the foot, however.] But as the blood in its return to the heart appears to pass almost entirely through the skin, which is thin and delicate, and also covered with cilia, there can be little doubt that the *whole surface of the body assists in aeration.*" After this, it seems to me that these authors need only reflect a moment on the absence of gill-structure in the papillæ, to perceive at once the impropriety of designating those organs as branchial, and of including the Eolis among Nudibranchiates. For although the papillæ expose a large surface to the air, they only do so in common with the rest of the skin ; mere extent of surface does not constitute a gill; nor will a merely respiratory surface constitute one. The frog, for example, respires by the skin, as well as by the lungs, but we do not call its skin a lung, because a lung has a specific structure, widely differing from the structure of the skin.

In their monograph, Messrs Alder and Hancock say: "The order Nudibranchiata is restricted to those animals bearing the character assigned to it by Cuvier; namely, the *possession of distinct external and uncovered gills.*" But we have just seen that it is precisely this character which is wanting in the Eolids; nor do I understand how these learned authors come to include Eolids among the Nudibranchiates, unless they also proposed to annihilate the very terms of their definition. If the Eolids have not distinct external and uncovered gills—if they have no gills at all, in fact—they cannot lawfully claim a place among the Naked-gilled Molluscs.

The reader will understand that at the coast, where these dissections were made, I had but a scanty supply of books to aid me, and that on returning home a diligent search was made, with a view of ascertaining what had been already noted on this point. Owen and Siebold assured me that I was right in denying the Eolids anything properly to be styled a gill. The former says: "In certain small shell-less marine genera—*e. g. Rhodope, Tergipes, Eolidina*—no distinct respiratory organs have been detected; these form the order *Apneusta.*" Professor Owen enters into no details; he simply asserts the fact, that no distinct organs exist. Siebold follows Kölliker, who establishes the order *Apneusta*, "in opposition to the other Gasteropoda, which have distinct respiratory organs." But the only reason adduced by him seems to me far from cogent; and I can understand how Alder and Hancock, who must have been perfectly aware of the position taken by Kölliker and Siebold, might altogether disregard it. Siebold says that the opinion respecting the branchial nature of the papillæ " is untenable, since it has been shown that they contain prolongations of the digestive canal."* I believe it to be untenable, simply because the papillæ contain *none* of the distinctive characters of gills; did they possess these, the fact of their also containing digestive prolongations would not deprive them of their rank as gills; any more than the fact of the heart, in some Molluscs, allowing the intestine to pass *through* it, deprives it of cardiac dignity.

Quitting this discussion for topics more surprising, let us fix our attention on one fact, cursorily indicated on a preceding page, namely, the existence, in the foot of the *Doris*, of a system of pores through which the water enters into the general cavity of the body. It is a fact which has not excited sufficient notice. Consider it, for a moment, and you will remark that water entering the general cavity must mingle with the blood, and that largely. But if it mingles thus freely with the blood, on its entrance, will it not also on its

---

* SIEBOLD – *Comp. Anat.*, English Trans., p. 249.

exit carry away a portion of the blood? It will, and does, as any one may prove experimentally. Agassiz has described it in detail, apropos of the *Mactra*,* and suggests that the pores of the fish may have a similar office. In the *Actiniæ* and *Acalephæ* the free passage of water to and fro is well known. In several Annelids (*Eunice, Nereis*, &c.), I have little doubt that the openings which exist on each side of the body, between every two feet, and which permit the passage of the ova into the surrounding water, are aquiferous tubes, through which the water passes to and fro. These openings were first described by Rathké; and Siebold admits that through them the water may probably pass. M. Quatrefages, however, a great authority on the subject of Annelids, is disposed to deny the existence of these openings: "J'ai vu chez l'Eunice sanguine des *apparences* rappelant un peu l'observation de Rathké; mais ces orifices, s'ils existent, ont évidemment pour objet de servir à la sortie des œufs; c'est là du moins ce que l'analogie permet d'admettre. Je n'ai jamais vu pondre ou éjaculer une Annelide errante."† What M. Quatrefages has not been fortunate enough to have seen, has several times been witnessed by me. I have seen the eggs issuing from these openings, the existence of which he is inclined to dispute; and inasmuch as these eggs may be seen floating in the general cavity of the body, before they issue through the openings at the side, there seems little room for doubt on the

* SIEBOLD ü. KÖLLIKER—*Zeitschrift. f. Wissen. Zoolog.* vii. 176.
† QUATREFAGES in *Annales des Sciences Nat.*, 1850; xiv. 298.

point. As to there being an habitual communication between the general cavity and the external water, through these openings, M. Quatrefages peremptorily denies it, without assigning his reasons. His mere assertion is of weight; but for the present I am strongly inclined to believe there *is* an habitual communication, because, besides ova, I have seen the diffusion of chyl-aqueous globules in the water, on the glass slide bearing a Nereid.

That an animal should suck up water into his body is intelligible enough; the water may serve for purposes of aeration, nutrition, locomotion, &c.: it may give the blood its oxygen, the tissues its organic substances in solution, the shell its salts, and it may serve as a fulcrum for the animal's progression. But when it is pressed *out* of the body—as in some animals it is with great frequency and rapidity—will there not be a spontaneous phlebotomy, of a rather dangerous nature? That is the question. Hereafter (Part III., Chap. II.) we shall have to consider more closely the relation of sea water to the circulating nutritive fluids; for the present let us be content with the idea that, in at least three divisions of the animal kingdom, water directly mingles with the blood, and quits it *en masse*, carrying some blood with it.

PART II.

SEA ANEMONES.

CHAPTER I.

SUDDEN ENTHUSIASM FOR THE ANEMONES—IMPERFECT STATE OF OUR KNOWLEDGE—LITERATURE OF THE SUBJECT—NO DISTINCTION BETWEEN ANIMAL AND PLANT: SENSITIVENESS, LOCOMOTION, AND CAPTURE OF FOOD—VORACITY OF THE ANEMONE—ITS PARALYSING POWER DENIED—HABITS AND INSTINCTS—IS IT VIVIPAROUS ONLY?—VARIATIONS OF COLOUR.

SINCE the British mind was all alive and trembling with the zoological fervour excited by the appearance of the hippopotamus in Regent's Park, no animal has touched it to such fine issues, and such exuberant enthusiasm, as the lovely Sea-Anemone, now the ornament of countless drawing-rooms, studies, and back parlours, as well as the delight of unnumbered amateurs. In glass tanks and elegant vases of various device, in finger-glasses and common tumblers, the lovely creature may be seen expanding its coronal of tentacles, on mimic rocks, amid mimic forests of algæ, in mimic oceans—of pump-water and certain mixtures of chlorides and carbonates, regulated by a "specific gravity test." Fairy fingers minister to its wants, removing dirt and slime from its body, feeding it with bits of limpet or raw beef; fingers, *not* of fairies, pull it about with the remorseless curiosity of science, and experiment on it, according to the suggestion of the moment. At once pet, ornament, and "subject for dissection," the Sea-Anemone has a well-established popularity in the British family-circle; having the advantage over the hippopotamus of being somewhat less expensive, and less troublesome, to keep. Were Sea-cows as plentiful as Anemones, one could not make pets of them with the same comfort. There would be objections to Potty in the drawing-room. There would be embarrassments in the commissariat. There would be insurgents among the domestics; for the best-tempered Jane might find it impossible to endure the presence of such a pet, and might resolutely refuse to bring up his water, and clean out his crib; whereas, although the red-armed Jane thinks you a little cracked when you introduce "them worm things" into your house, she keeps her opinions within the circle of the kitchen, and consents to receive her wages without a murmur.

It is difficult to say what occasioned this sudden enthusiasm for Anemones: lovely, indeed, but by no means the most lovely, and certainly not the most interesting wonders of the deep. Mr Gosse by his pleasant books, and Mr Mitchell by his tanks in the Regent's Park Zoological Gardens, have mainly contributed to the diffusion of the enthusiasm; and now that enterprise has made a commercial branch of it, we may consider the taste established, for at least some years.

One good result of this diffusion will be an extension of our knowledge, not only of these, but of many other of the simpler animals. For several years the writings of zoologists have given a place to observations on the Anemones; but the observations have been incomplete, and all handbooks and treatises which repeat these observations are very naturally crowded with errors. To give the reader an idea of the state of current opinion on this one topic, it is enough to mention that in one admirable handbook, the second page, devoted to a description of the habits of the Anemone, contains six distinct errors: yet this is no fault of the compiler; he states what preceding writers state, and his excellent summary of what is known bears the date of 1855. If the habits have been so imperfectly observed, you may guess what a chaos the anatomy and physiology of this animal present. Such being the state of the case, we may hope that the wide diffusion of a taste for vivaria will in a little while furnish Science with ample material; and meanwhile, as many of my readers are possessors of vivaria, actual or potential, and will certainly not content themselves with blank wonderment, but will do their utmost to rightly understand the Anemones, even if they make no wider incursions on the domains of the zoologist, they will perhaps be interested if I group together the results of investigations, pursued at Ilfracombe and Tenby during the summer of 1856, and, with less energy, because with less prodigality of specimens, during the autumn and winter at home; adding thereto some corrections derived during the spring and summer of 1857. In the present state of knowledge, the independent observations of every one who has had any experience cannot but be welcome.

The literature of the subject is extensive if we include all the passing notices made by naturalists from Aristotle downwards; but the capital works are few. What Aristotle says of them is accurate enough for the most part,* although the details are scanty. Owing to his commentators being less accurately informed than he was, they have misunderstood the passage wherein he speaks of the sea-nettles quitting their rocks in quest of food at night, and have supposed that he alluded to the Medusæ. But by ἀκαλήφη he indubitably meant the Actiniæ as well as the Medusæ; and he was right in saying they sometimes quitted the rocks to which they had fixed themselves. Rondelet adopted Aristotle's term of *Urticæ marinæ*, adding to it the epithet of *adfixæ*, to distinguish them from the errant Medusæ. Réaumur in 1710 began to investigate them more seriously than any of his predecessors had done; but the Abbé Dicquemare must be counted as the first good authority on the subject. He furnished the most extensive and reliable information, in three papers of the *Philosophical Transactions* for the years 1773, 1775, and 1777. The name of *Actinia* was adopted by Linnæus from Patrick Brown, and has since been universally accepted. The reader will understand, therefore, that Sea-Anemone and Actinia are convert-

---

* ARISTOTLE: *Hist. Animal.* lib. i. c. i. 8, and lib. iv. c. vi., 4. There are probably other passages, but these are all I have been able to find.

ible terms, and may be employed indifferently. In 1809, Spix published an account of the anatomy of the Actinia, which was, however, very faulty; its announcement of the discovery of a nervous system has long been rejected. It was not even accepted by Delle Chiaje, the next writer of authority, whose *Memorie sulla storia e notomia degli animali invertebrati del regno di Napoli*, 1822–29,* is still quoted by systematic writers. There are some errors in this work; among them is that of Spix, who imagined he had discovered a vascular system; but in spite of inaccuracies, Delle Chiaje's account is worth studying. The next, and up to that time the most important publication, was that of Professor Rapp,† which may still be read with profit, and was only displaced by Dr Johnston's elaborate *History of British Zoophytes;* and this in turn will have to give way to Mr Gosse's "Monograph on Sea-Anemones," when that much-expected work makes its appearance;‡ meanwhile his *Rambles of a Naturalist*, and his *Tenby*, contain valuable

---

* I have not seen this book, but I found the magnificent work, *Descrizione e notomia degli animali invertebrati della Sicilia Citeriore* (published in 8 vols. folio; Naples, 1841-44), on the shelves of Mr Trübner in Paternoster Row. It was too costly for an author's purse; but Mr Trübner, with characteristic generosity, insisted on lending it to me for some weeks; the offer was too tempting to be refused. It is a book to make a naturalist languish with desire.

† *Ueber die Polypen im Allgemeinen und die Actinien ins besondere*, 1829. This work contains three coloured plates, which were thought admirable in those days, but which our progress in the art of illustration throws into the shade.

‡ Since published.

notices. Mr Tugwell's *Manual of the Sea-Anemones usually found on the English Coast* is specially addressed to amateurs, and contains useful information pleasantly conveyed, and coloured plates of rare excellence. Of a rigorously scientific character are the two memoirs published by M. Hollard, one "On the anatomy of the Actinia,"* and the other *Études Zoologiques sur le genre Actinia*.† Nor, although I have not been able to procure it, should Mr Teale's paper on the Anatomy of the *Act. coriacea* in the "Leeds Philosophical Transactions" be forgotten, since it has formed the authority for subsequent writers. Separate points have been treated by Erdl, Quatrefages, Wagner, Kölliker, Leuckart, and others; but the five works just mentioned are, I believe, all the reliable original sources of information on the structure and habits of the Sea-Anemones; and they all contradict each other with great freedom, so that the student need not be surprised if he, in turn, is forced to oppose a flat denial to many a positive assertion. In the course of the present pages such flat denials will be frequent.

It must be assumed at starting that the reader knows what a Sea-Anemone is, in aspect at least. No description will avail, in default of direct observation; even pictures only give an approximate idea; while to those who have seen neither picture nor animal it will be of little use to declare that the "Actinia is a fleshy cylinder, attached by one extremity to a

---

* *Annales des Sciences Nat.*, 1851, vol. xv.
† *Revue et Magazin de Zoologie*, 1854. No. 4.

rock, while the free end is surmounted by numerous tentacula arranged in several rows, which, when expanded, give the animal the appearance of a flower." Assuming, then, that you know the general aspect of the Actinia, you may follow my description of the animal's bearing and habits.

How do I know that it is an animal, and not a flower, which it so much resembles? No one yet has been able to distinguish, in the face of severe critical precision, between the animal and plant-organisation, so as to be able authoritatively to say, "This is exclusively animal." To distinguish a cow from a cucumber requires, indeed, no profound inauguration into biological mysteries; we can "venture fearlessly to assert" (with that utterly uncalled-for temerity exhibited by bad writers in cases when *no* peril whatever is hanging over the assertion) that the cow and cucumber are not allied—no common parentage links them together, even through remote relationship; but to say *what* is an *animal*, presupposes a knowledge of what is essentially and exclusively animal; and this knowledge unhappily has never yet been reached. Much hot, and not wise, discussion has occupied the hours of philosophers in trying to map out the distinct confines of the animal and vegetable kingdoms, when all the while Nature knows of no such demarcating lines. *The Animal does not exist; nor does the Vegetable:* both words are abstractions, general terms, such as Virtue, Goodness, Colour, used to designate certain groups of particulars, but having only a mental existence. Who has been fortunate enough to see the Animal? We have seen cows, cats, jackasses, and camelopards; but the "rare monster" Animal is visible in no menagerie. If you are tempted to call this metaphysical trifling, I beg you to read the discussions published on the vegetable or animal nature of Diatomaceæ, Volvocinæ, &c., or to attend to what is said in any text-book on the distinctions between animals and vegetables; and you will then see there is something more than metaphysics in the paradox. In the simpler organisms there is *no* mark which can absolutely distinguish the animal from the vegetable; and if in the higher organisms a greater amount of characteristic differences may be traced, so that we may, for purposes of convenience, consider a certain group of indications as entitling the object to be classed under the Animal division, we must never forget that such classifications are purely arbitrary, and as the philosophers say—*subjective.*

Now what are the characteristic marks of the Sea-Anemone, which entitle it to be removed from the hands of the botanist, and placed in those of the zoologist? Rymer Jones declares that its animal nature " is soon rendered evident," this evidence being the manifestation of sensibility. "A cloud veiling the sun will cause their tentacles to fold as though apprehensive of danger from the passing shadows." Unhappily, the fact alleged is a pure fiction; and, were it true, would not distinguish the Actiniæ from those plants which close their petals in the dark. A fiction, however, it is, as any one may verify. If Actiniæ have been seen to fold up their tentacles when a cloud has passed before the sun, this has been a coincidence, not a causal relation; so far from light being the necessary condition of their expansion, they are in perfect expansion in the darkness; and if the venturous naturalist will, with the solemn chimes of midnight as accompaniment, take his lantern on the rocks, he will find all the Anemones in full blossom. I said that Rapp might still be read with profit. Hear him on this point. "Many Zoophytes, although without eyes, possess the power of distinguishing light from darkness. This has long been asserted of the Actiniæ, nevertheless, on some species, it appeared to me that neither light nor darkness exercised any appreciable influence. *Actinia plumosa*, which I often watched on the western coast of Norway, expanded its tentacles equally in the dark, as when I removed it suddenly from darkness into direct sunlight. The *Actinia depressa* suddenly collapsed when direct sunlight was allowed to fall on it." I have repeatedly tried the experiment of overshadowing the pool in which the Actiniæ were expanded, but never saw them retract their tentacles in consequence. The reader will not suppose that I mean to deny the sensitiveness of the Actiniæ. I am merely answering an argument which several writers have repeated respecting the alarm felt by the Actiniæ when a cloud veils the sun, or a shadow affects them.

But the Anemone must be an animal, you suggest, because it is seen to catch and swallow other animals.

This, however, is no proof. Although the Anemone entraps its prey, or anything else that may come in contact with its tentacles, this is no proof of animality; the sensitive plant, known as the Flytrap of Venus (*Dionæa muscipula*), has a precisely analogous power; any insect, touching the sensitive hairs on the surface of its leaf, instantly causes the leaf to shut up and enclose the insect, as in a trap; nor is this all: a mucilaginous secretion acts like a gastric juice on the captive, digests it, and renders it assimilable by the plant, which thus feeds on the victim, as the Actinia feeds on the Annelid or Crustacean it may entrap. Where, then, is the difference? Neither *seeks* its food: place the food within a line's breadth of the tentacles, or of the sensitive hairs, and so long as actual contact is avoided, the grasping of the food will not take place. But you object, perhaps, that this mode of feeding is normal with the Actinia, exceptional with the Flytrap. The plant, you say, is nourished by the earth and air, the animal depends on what it can secure. Not so. For granting—what, in fact, I sturdily dispute—that the Flytrap is in no way dependent upon such insect food as may fall into its clutch, we shall still observe the Actinia in similar independence. Keep the water free from all visible food, and the Actiniæ continue to flourish and propagate just as if they daily clutched an unhappy worm. The fact is well known, and is currently, but erroneously, adduced as illustration of the animal's power of fasting. But there is no fasting in the matter. In this water free from visible aliment there is abundance of invisible aliment,—infusoria, spores, organic particles, &c., which the animal assimilates, much in the same way as plants assimilate the organic material diffused through the soil and atmosphere. Filter the water carefully, and remove from it all growing vegetation, and you will find the animal speedily dying, however freely oxygen may be supplied. It is on this account that when we make artificial sea-water, it is necessary to allow algæ to grow in it for some two or three weeks before putting in the animals; the water becomes charged with organic material.

Mere sensibility and capture of food, therefore, are not the distinguishing marks we seek, since the plant is found to possess them as perfectly as the animal. Is spontaneous locomotion a sufficient mark? No; and for these two reasons: Some animals have *no* such power; some plants, and all spores, *have* it. There are animals which no botanist has ever claimed—the Ascidians, for example, which can scarcely be said to exhibit any motion at all (the rhythmic contraction and expansion of their orifices not deserving the name); their whole lives are spent rooted to the rock or shell, as firmly as the plant is rooted in the earth. Nay, even with regard to the Anemones, it is said by Dr Landsborough, Dr Carpenter, and others, that they will not move towards the water, should the vessel be gradually emptied, or the water evaporate, not even if their tentacles can reach its surface. This is incorrect; but I mention it as one of the difficulties which would meet the student in the way of distinguishing the

Anemone from plants. It is one of the many inaccurate statements grounded on imperfect observation, which are repeated in handbooks. The original observer probably noticed an Anemone some time out of the water, making no effort to return; had the observation been continued, the doubt would have been solved. Some Anemones, especially the Common Smooth species (*Mesembryanthemum*), are accustomed daily to be left out of water by the receding tide, so that in captivity they may be supposed rather to enjoy an occasional air-bath. I have repeatedly seen mine crawl out of the water and settle on the edge of the glass, or pan, high and dry; but they descended again after a few hours. The locomotion of the Anemones is, however, various in various species. I do not think the "Trogs" ever move; nor do the "Gems" seem migratory; but the "Antheas" and the "Smooths" are somewhat restless. "The Actiniæ," says Rymer Jones, "possess the power of changing their position; they often elongate their bodies, and, remaining fixed by the base, stretch from side to side, as if seeking food at a distance; they can even change their place by gliding upon the disc that supports them, or detaching themselves entirely, and swelling themselves with water, they become nearly of the same specific gravity as the element they inhabit, and the least agitation is sufficient to drive them elsewhere. Réaumur even asserts, that they can turn themselves so as to use their tentacles as feet, crawling upon the bottom of the sea; but this mode of progression has not been observed by subsequent naturalists." Yes, Dr Johnston once saw it; I also witnessed an Anthea moving thus; but I suspect it is only the Anthea which has the power, and this it probably owes to its more solid tentacles.*

Again the question recurs, How then do we know the Anemone to be an animal?—in other words, what characteristic marks guide zoologists in classing it in that division? I really know of none but purely anatomical marks.† These, however, suffice, and if you please we will continue to speak of the Anemone as an animal, and, what is more, a very carnivorous animal, eating most things that come within reach, from limpets to worms, from fish to roast beef. It has even a reputation for voracity, not to say *gourmandise*; in the matter of shell-fish it would put even Dando to the blush. Dr Johnston, in his valuable *History of British Zoophytes*, relates this anecdote: " I had once brought

---

* Rapp, *loc. cit.* p. 44, says he has often witnessed it; but he only mentions the Anthea as possessing the power. Aristotle, as we have seen, makes special mention of it.

† It is unnecessary to particularise these anatomical marks, which will occur to the mind of every student, as belonging exclusively to that division of animated beings which manifest the group of phenomena baptised by the name of Animality. Wherever you find muscular tissue, or an alimentary canal, you are absolutely certain that nothing belonging to what finds its place in the group of marks which indicate the vegetable kingdom, is before you. In function there is often considerable resemblance between Plant and Animal; but in structure the differences early manifest themselves, growing greater as the scale ascends. Although, therefore, at the bottom of the scale no distinguished characteristic isolates animals from plants, as we ascend the scale we find many definite marks by which the two groups may be known.

to me a specimen of *Actinia crassicornis* that might originally have been two inches in diameter, and that had somehow contrived to swallow a valve of *Pecten maximus* of the size of an ordinary saucer. The shell fixed within the stomach was so placed as to divide it completely into two halves, so that the body stretched tensely over had become thin and flattened like a pancake. All communication between the inferior portion of the stomach and the mouth was of course prevented; yet instead of emaciating and dying of an atrophy, the animal had availed itself of what had undoubtedly been a very untoward accident, to increase its enjoyments and its chance of double fare. A new mouth furnished with two rows of numerous tentacles was opened upon what had been the base, and led to the under stomach —the individual had become a sort of Siamese Twin, but with greater intimacy and extent in its unions." Such is the blind voracity of this animal, that anything and everything is carried straightway into its stomach to be there tried, and rejected only on proved incompatibility.

One day, while sorting and distributing to their respective jars the animals captured during the morning's hunt, I was called into the balcony by the agitated entreaties of lovely Sixteen, exclaiming, "Oh, do come! do come, and rescue this green Anemone from a great nasty beetle." I went to the rescue, and found a large beetle struggling in the clutches of a green Anthea. "The beetle is the victim," I quietly told Sixteen; who, not having profound sympathies with beetles, was pacified as she saw the struggling insect slowly passing into the stomach of the Anthea, his struggles growing fainter and fainter, and finally ceasing altogether, till at last we saw him with head and thorax engulfed in the ravenous maw, his abdomen sticking up in the air.

A question of great interest and some intricacy here presents itself: Was the beetle *paralysed* by some peculiar poison secreted from the tentacles of the Anemone?—a question which opens into this wider one: Have the Polypes the mysterious power, almost universally attributed to them, of paralysing with a touch the victims they may grasp, so that, should the victim escape from their grasp, it is only to die presently from the fatal touch? The power of fascination possessed by some animals, of poisoning possessed by others, of electrical discharges possessed by others, naturally lead men to interpret certain observations made on the Polypes, as proofs that they, too, possess some such power: and this suggestion gains a more ready credence from the tendency in most minds to welcome every unexplained phenomenon as indicating an occult cause. This witch-like power of fascination,—this power of paralysing with a touch, appeals to our imagination, and gains easy access to belief. But the spirit of scientific scepticism forces me to declare that, as far as my observations and experiments extend, there is nothing like evidence in favour of this power, much evidence against it. Some Anemones certainly appear to sting—as some jellyfish sting—although the majority have no such effect upon our hands, which every one

knows who has handled them. I never perceived this stinging sensation myself; and Dr Landsborough says: "From my own experience I can say nothing as to this stinging power; for though I have handled not only the commoner Actiniæ, but also the larger and less common Anthea, I never felt anything approaching to stinging; but I never touched a tentaculum without perceiving the tip of it had some prehensile property by which it took a slight hold of the skin of the finger, causing a kind of rasping feeling when withdrawn. It may be, however, that the fangs had not fair play with my fingers, if somehow or other they are sting-proof."*
He then makes the following quotation from Mrs Pratt's *Chapters on the Common Things of the Sea-side*, which I reproduce as positive and direct testimony: "It appears that different persons are variously affected even by touching the same Actiniæ. The author had placed in a vessel of sea water a fine specimen of the fig marygold sea-anemone, which she was accustomed to touch many times during the day. The tentacula closed immediately round the intruding finger, producing only a slight tingling. Her surprise was great at finding that the same anemone, on being touched by another person, communicated a more powerful sensation, which her friend assured her was felt up the whole of the arm. More than twenty persons touched this Anemone; and the writer was amused by observing how variously they were affected, some being only slightly tingled, while others started

---

* *Popular History of British Zoophytes*, p. 239.

back as if stung by a nettle." I think, in the face of testimony so precise as this, we may waive all negative evidence, and accept the fact of stinging as proven.*

But now comes the question: Is this stinging produced by poison-vesicles and spicula, as the great majority of writers maintain; or is it no more poisonous than the pricking of a thorn? Those who maintain the former opinion, explain by it the alleged cases of paralysis exhibited by the animals which have escaped in the struggle; and the incident just related of the beetle killed, but not swallowed, seems entirely to favour such a conclusion. Nevertheless, from subsequent investigations, I am led to oppose the opinion *in toto*. Sir John Dalyell—one of the best authorities—thinks that the Anemone conquers its prey by mere strength, and not by any poisonous fluid. He is somewhat exaggerated, however, in the statement of his opinion. "Nothing," he says, "can escape their deadly touch. Every animated being that comes in slightest contact is instantly caught, retained, and mercilessly devoured." This is mere rhetoric: animals, even such as form their natural prey, constantly touch the tentacles—nay, are even caught, and yet escape. "Neither strength nor size, nor the resistance of the victim, can daunt the ravenous captor. It will readily grasp an animal which, if endowed with similar strength, advantage, and resolution, could certainly rend its body asunder. It is in the highest degree carnivorous. Thence do all the varieties

---

* ARISTOTLE, lib. iv. c. vi. 4, mentions their stinging, οὕτως ὥστε τὴν σάρκα ἐπανοιδεῖν.

of the smaller finny tribes, the fiercest of the crustacea, the whole vermicular race, and the softer tenants among the testacea, fall a prey to the Actiniæ." One is astonished to meet with such a passage from so accurate an observer. It is pure exaggeration, which succeeding writers have accepted as literal truth. Thus, Rymer Jones says, that "no sooner are the tentacles touched by a passing animal, than it is seized and held with unfailing pertinacity." Had the professor given his attention to Anemones he would know that, so far from the grasp being "unfailing," it as often fails as succeeds, when the captive is of tolerable activity; and very noticeable is the fact, that when the animals escape, they escape *unhurt;* a fact in direct contradiction to the belief in a poison secreted by the tentacles.

I resolved to bring this question to the test, and dropped a tiny crab, rather smaller than a fourpenny piece, on the tentacles of my largest Crassicornis (nearly as large as a glass tumbler). He was clutched at once, and the tentacles began to close round him; he struggled vigorously, and freed himself after a few seconds. Placed there a second time, he again got away. I waited to see if any symptoms of paralysis would declare themselves after this contact, but he was as lively as ever. Later in the day I placed him on the tentacles of the voracious Anthea, the most powerful of all the Anemones, and the only one which seems to sting; but the crab was too active, or too little appetising; he got away as before. I tried another Anthea and a Daisy (*Actinia bellis*), but with the same result. In each case the crab was clutched, but in each case he got away unhurt. I then chose another crab, not more than half the size of the former, and certainly no match in point of strength for the Anemone, yet after being embraced and carried to the mouth, I observed the crab slowly appear from the unfolding tentacles, and scuttle away with great activity.

This experiment casts a doubt on what is asserted by all writers, namely, that Anemones feed on crabs. Rymer Jones records that "they will devour a crab as large as a hen's egg." Has any one ever *seen* a live crab caught and eaten by an Anemone? I confess never to have seen it, and the experiment just related disposes me to doubt: although it is quite possible that my Anemones were dainty, because not hungry, and refused food which, under less epicurean conditions, would have been welcome. If any one has *seen* the Anemone feeding on live crabs, that is enough. Meanwhile I think it right to propound the doubt, and to add to it this subsequent observation: I took a tiny Crustacean, of the shrimp family, about half an inch in length, and dropped it in a vase containing some Daisies. It soon touched the tentacles of one of these, was drawn in, but almost immediately escaped. It then swam about until it touched the largest Daisy, and was quickly engulfed. As it had entirely disappeared, I expected it would be certainly killed if not eaten, but in a few moments it made its way out unhurt, and swam away. These Daisies had not been fed for at least a fortnight; they had subsisted entirely on the invisible aliment floating in the water: yet they either could not, or would not, eat this Crustacean.

No one can have taken Anemones from the rocks without observing fragments of small crabs, and sometimes whole crabs, as big as crown pieces, in their stomachs; but the question whether these crabs were captured alive by the Anemones is not thereby answered. Without absolutely denying that the Anemone does thus capture them, I am forced by repeated observation and experiment, to declare that the evidence all points in the direction of a denial. I remember once accidentally dropping a tiny crab on the expanded oral disc of a Crassicornis, whose mouth was wide open; and very amusing it was to see the little creature rush into the open mouth, settle himself comfortably there, and begin twittering his antennæ, as crabs do when their alarm subsides. The *Crass* never moved: owing to the insensibility of the stomach, he was quite undisturbed by this refugee—in this by no means resembling the *Cyclops* of Euripides, who energetically repudiates the idea of swallowing the satyr:

> "You in my stomach? Horror, if I had!
> Your capering antics there would drive me mad." *

The crab remained twittering for some minutes. I touched him, and he retreated deeper down into the cavity. Looking into the pan some time after, I found he had crawled away, so I gave him to an Anthea,

---

* ἥκιστ'· ἐπεὶ μ' ἂν ἐν μέσῃ τῇ γαστέρι
πηδῶντες ἀπολέσαιτ' ἂν ὑπὸ τῶν σχημάτων.
— *Cyclops*, v. 220.

who clutched, but soon released him. I then gave him to another *Crass*, who swallowed him, but in a little while slowly unclosed his tentacles and let him escape; being apparently of the Cyclopean opinion that the capering antics would produce a gastric fever. Now, it is quite possible that the Anemone, having clutched the carcass of a dead crab washed on to its tentacles, and having swallowed it, as it swallows most objects once clutched, may retain the carcass in its stomach, and quietly digest the flesh thereof, as far as digestion is possible,* yet be unwilling to retain the live crab under similar circumstances, because of the incessant struggles of the victim; and I direct the attention of students to the point, because if any one witnesses the capture and digestion, it will be enough to destroy all negative evidence.

On the question of food we may withhold our opinion till some more decisive evidence is adduced; but on the question of the paralysing power said to reside in the tentacles, these experiments surely determine a negative. In spite of the beetle, so completely vanquished, there is the evidence of crabs and shrimps being in repeated contact with the tentacles, and in nowise affected.

While preparing these notes for the press, I have been led to extend the experiments; because, although it would by no means necessarily follow that whatever was true of the Hydroid Polypes must also be true of

---

* In Part III., Chap. I., the reader will see that digestion, in the strict sense of the word, is *not* possible.

the Anemones, yet a very plausible suspicion might arise—and did indeed arise in my mind—throwing doubt on results which were in contradiction to what was reported of the fresh-water Polypes. Read this passage from the last edition of Owen's *Lectures*, bearing the date 1855: "That the tentacula have the power of communicating some benumbing or noxious influence to the living animals which constitute the food of the hydra, is evident from the effect produced, for example, upon an entomostracan, which may have been touched, but not seized, by one of these organs. The little active crustacean is arrested in the midst of its rapid darting motion, and sinks apparently lifeless for some distance; then slowly recovers itself, and resumes its ordinary movement. Siebold states, that when a Naïs, a Daphnia, or the larva of a Cheironomus, have been wounded by the darts, they do not recover, but die. These and other active inhabitants of fresh waters, whose powers should be equivalent to rend asunder the delicate gelatinous arms of their low-organised captor, seem paralysed almost immediately after they have been seized, and so countenance the opinion of Corda, that the secretion of a poison enters the wounds." Such statements can only be set aside by direct experiment; and the superiority of *Experiment* over mere Observation needs no argument.

As a matter of observation, I too had been struck with the fact noticed by Owen. I saw the tiny Water-fleas drop apparently lifeless to the bottom of the phial, after being some time held by the tentacle of the Hydra; and, intently watching them, I saw them at last swim away again as lively as before. I removed a Hydra from the phial, in a little water, and placing it on a slip of glass, allowed it to settle and expand there for two hours, when I added several water-fleas (*Cypridæ*) to the little pond, and patiently watched them swimming to and fro. Repeatedly they touched the tentacles in their course, but were not hurt, were even not arrested. At length one was caught, and held for some seconds; it then fell to the bottom, and remained motionless for at least two minutes, after which it started up, and was off as if its course had never been arrested. This certainly had very much the appearance of a case of slight paralysis; the animal seemed arrested by some benumbing influence, which for two minutes rendered it powerless; at the expiration of that time it seemed to have sufficiently recovered itself to swim away. If Observation alone sufficed, in questions so complex as those of Biology, this observation would have confirmed the statements of Siebold, Corda, and Owen. But observation alone does not suffice. I bethought me of a simple experiment. With a needle I gently arrested one of these water-fleas; it suddenly sank motionless, remained thus for more than a minute, and then darted off again. Thrice I repeated this act, and each time with similar result. Will any one say the needle had a paralysing power, or a benumbing poison which was secreted, when the animal came in contact with it? And does not the reader at once recognise in this sudden motionlessness of the animal a very familiar phenomenon? The spider, the crab, the oniscus, and very many animals "sham dead," as schoolboys know, when danger threatens; these water-fleas "sham dead" when the Polype or the needle touches them. I might have rested my incredulity of the alleged paralysing influence on this one experiment; but I confirmed it in other ways. Dropping the larva of an Ephemeron into the phial containing the Hydræ, I observed it thrice caught by three different Hydræ; it did not "sham dead," but tore itself away without visible hurt. Nay, I also observed one of those animalcules known as "paste-eels" for some time in contact with the tentacle of a Hydra, on the stage of the microscope, but, in spite of its having no shell to protect it from the poison, it was unhurt by the contact. Not having a Naïs, I could not test what Siebold says of it; but what has already been mentioned must, I think, suffice to convince the reader that the current opinion is an error, founded on Observation unverified by Experiment. It was only by verification, according to the demands of inductive scepticism, that the error became obvious.*

Had Trembley's celebrated work on the fresh-water Polypes been read by one in a hundred of those who

---

* The day this was written I could not rest till I had dredged a favourite pond, and brought home a supply of Naïds, with which, on the following morning, I tested Siebold's statement. First, I placed a *Naïs filiformis* in a glass cell with a *Hydra viridis;* but although its wriggling constantly brought it into contact with the tentacles, it was never grasped. I then placed a Naïs in the phial containing many

cite it, the error just noticed would never have gained currency; the observations of that very accurate observer would have suggested a fallacy in the interpretation. " J'ai vu souvent des Pucerons (water-fleas) qui parvenoient à se mettre en liberté. Il m'a paru qu'ils le pouvoient plus facilement que les Millepieds (Naïds). Comme ils sont fort petits, et surtout que leurs corps n'est pas allongé, ils risquent moins que les vers de s'engager dans les bras des Polypes en se débattant."* And he nowhere mentions that the escaped animals died, or gave any signs of having been paralysed by contact with the tentacles.

To sum up this discussion, we may say that ample testimony is afforded in proof of the position that *Anemones have a certain stinging power*, the nature of which is not yet ascertained, but which is probably a minor degree of that possessed by some Jellyfish.

Direct experiment and observation prove that *neither the Anemones, nor the fresh-water Polypes, possess in any degree the power of paralysing other animals.*

" But do tell us something about the habits and instincts of these Anemones," some light-minded reader

---

Hydræ; it was instantly caught by one, and held for some time till it struggled itself free. Not only was it apparently unhurt by this contact, but to-day it is as lively as it was three days ago, just before the experiment. With two other Naïds the same result was observed. This completes the overthrow of the current opinion respecting the Hydra's paralysing power.

* TREMBLEY: *Mémoires pour servir à l'hist. des Polypes d'eau douce.* 1744, p. 92.

K

suggests, impatient of all discussion, and supremely indifferent to all considerations save those of a moral order. Unhappily my story is not ampler in detail, nor finer in complexity of movement, than the story of Canning's " Knife-grinder "—who had none to tell. The Anemone is lovely, but even its warmest admirers must confess it is a little monotonous in its manifestations. Existence suffices it. It expands its coronal of tentacles, eats when chance favours it, produces offspring, which it sends forth, leaving them, borne by the many currents of the sea, to settle where they list, without any fear of parental supervision, and thus lives to a good old age, if no one nudges the elbow of Atropos, and causes that grim lady suddenly to cut the thread.* Nor is it easy to nudge the old lady's elbow. The Anemone has more lives than a cat. We have already seen (p. 21) how it will resist slashing and amputation ; and, except zoologists and the fierce little *Eolis*, I know no animal which, finding its flavour agreeable, cares to make a meal of it. Yet, curiously enough, this Anemone, whose vitality is so remarkable, who may be hacked and hewed without appearing to suffer from it, dies almost immediately if removed from salt water to fresh. It will live for days out of water, but in fresh water it will not live at all. There

---

* The age to which an Actinia may live has not yet been definitely ascertained ; but Professor Fleming at Edinburgh has [Oct. 1857] one in his possession, which was taken at North Berwick in 1828 ; so that, at the very least, it must be twenty-eight years old, that period having been passed in confinement.

is a problem for the physiologist. He will probably see in it the effect of endosmosis, by which all the fluids in the tissue of the Actinia are suddenly so diluted as to be rendered unfit for the vital processes.

The Anemone has little more than beauty to recommend it; the indications of intelligence being of by no means a powerful order. What then? Is beauty nothing? Beauty is the subtle charm which draws us from the side of the enlightened Miss Crosser to that of the lovely though "quite unintellectual" Caroline, whose conversation is not of a brilliant kind; whereas Miss Crosser has read a whole Encyclopædia, and is so obliging as to retail many pages of it freely in her conversation.

Besides, if the monotony of the Anemone wearies you, there is always this variety in reserve: you can eat it! The Italians do; they boil it in sea water with great satisfaction. Thus boiled, it has "a shivering texture, somewhat like calf's-foot jelly; the smell is somewhat like that of a warm crab or lobster," and it is eaten with savoury sauce. Mr Gosse describes his frying them in butter, if I remember rightly; and although he felt a little difficulty in swallowing the first mouthful—probably remorse, and zoological tenderness, gave him what the Italians call a "knot in the throat"—yet, having vanquished his scruples, he ate with some relish. Lady Jane is "horrified" at the idea of eating her pets; but now that horse-flesh is publicly sold in the markets of Vienna and other German towns, and public banquets of hippophagists are frequent in France,* will Anemones long escape the frying-pan?

It was hinted just now that the Anemone was but an indifferent parent. Having given birth to her offspring, she spends no anxious hours over the episodes of infancy. When I say She, I might as well say He, or It, for no distinction of sex exists, as we shall see presently; and probably it is to this cause that the parental indifference may be traced; how, indeed, can maternal tenderness and ceaseless vigilance be expected, when the maternal individual is as yet undeveloped? The Actiniæ are viviparous. Indeed I suspect they are *only* viviparous, and not at all oviparous. Rymer Jones seems to hesitate on the point, adding, "but it is asserted by numerous authorities that the young are not unfrequently born alive." I not only assert this, but ask whether any one has ever seen the contrary? It startled me, however, when, on opening an Anemone, I for the first time saw a young one drop out, and immediately expand its tentacles; and some days afterwards, as I was carrying home a lovely "Gem," I saw first one, then two, three, four, seven young ones issue from its mouth, fix themselves at the bottom of the vase, and make themselves at home; they were of various sizes, and in various stages of development. Since then, I have repeatedly witnessed this mode of birth; and one day, seeing something in the inside of the tentacle of a Daisy, I snipped the tentacle off, and found a young Daisy there. Some writers imagine that the young

* See *Physiology of Common Life*, vol. i. p. 162.

issue through orifices at the tips of the tentacles—a supposition not very credible. The truth is, that at the bottom of the stomach there is a large opening—not several minute openings as we see figured in books—through which the young pass from the general cavity into the water; and this appears to me the only exit for the young. Without absolutely denying that the ova are extruded, and their early development carried on out of the parent's body, I have never been able to detect ova, except within the parent.

I leave this passage as it originally appeared, although, at Jersey, I was subsequently convinced that, with respect to one species at least, the doubt expressed should be withdrawn. In the water of a pan containing, among other animals, specimens of *Actinia parasitica*, I twice noticed abundance of light-purple ova floating at the surface. Some of these were placed in a vase by themselves, and others left in the pan; but no further development took place. One day dissecting a *Parasitica*, I found in its ovaries the same kind of purple ova. This seems very like evidence that the *Parasitica*, if no other species, is oviparous; and it is strengthened by the fact that, as far as my experience extends—and I have had scores of specimens—the *Parasitica* is not viviparous. The point needs elucidation, and the student may amuse himself with it—first by endeavouring to prove the other species to be really oviparous, as well as viviparous; secondly, by ascertaining whether the *Parasitica* is viviparous.

In the visceral cavity of a smooth Anemone a young one as large as a cherry was found; and to complete the marvel, it was faintly striped with green, like the well-known "green-striped variety," although its parent was of a dark-brown hue. Could the old one have swallowed an errant youth by mistake? No. It had been many weeks in captivity, where no such errant youths were within reach: besides, Anemones do not swallow each other, or if they should, *par distraction*, make a mouthful of a friend, they would quickly throw him up again: cannibalism belongs to a higher grade of social development. Apropos of this peculiarity of colour, I may remark on the great variations observable in the colour of Anemones, and the impropriety of making colour the distinguishing mark of species. Thus, to select a striking example, Mr Gosse makes two distinct species of the Orange-disked and Orange-tentacled Anemones, naming them Venusta and Aurora; but as if to prove the indifference of all such characteristics, I brought with me from Tenby an Orange-disked —and only one—which, before it had been home a fortnight, I discovered, with great surprise, was changed into an Orange-tentacled—disc and tentacles being of a rich orange hue, the only traces of white remaining just at the tips. If there had been any other specimen in the vase I might have doubted; but having only one in company with a white Daisy, and a Smooth Anemone, there was no avoiding the conclusion.

I have had an Anthea, with brilliant green tentacles, turn to a pale grey in the course of two days, and back again to green: and a Weymouth Anemone turn from pearly-white to a soft reddish-brown. In fact the changes of colour, except in the Crassicornis, which appears to me to retain its hues with tolerable constancy, are far too frequent to admit of colour forming a specific character.

CHAPTER II.

DESIRES FOR ABUNDANCE—THE THREAD-CAPSULES: ARE THEY NETTLING ORGANS?—FUTILITY OF OBSERVATION UNCONTROLLED BY EXPERIMENT—STRUCTURE OF AN ANEMONE—GENERAL LAW OF DEVELOPMENT—DEVELOPMENT OF THE HUMAN HAND—REPRODUCTION OF ANEMONES—THEIR OVARIES AND SPERMATOZOA—ARE ANEMONES OF SEPARATE SEXES?

CHARLES LAMB, in one of his exquisitely humorous letters, remembering the prodigal command of paper which he enjoyed as a clerk in the India House, and comparing it with his forced stinginess in that article now he is no longer clerk, refers to the probable feelings of Adam when purchasing a pennyworth of apples, "from an applewoman's stall in Mesopotamia," and recalling the prodigal abundance of Paradise. Dr Johnson said that never but once in his life had he found himself possessor of as much wall-fruit as he could eat. These two lingering retrospects of former plenty appeal to me forcibly: it is true that in the particular case of apples, a matured taste, fortified by philosophy and modified by dyspepsia, renders one tolerably resigned to poverty—and in the case of wall-fruit, the reader, terrified by absurd rumours as to the cholera-influences supposed inevitably to issue from plums, peaches, apricots, and nectarines, may be inclined to consider a limitation of quantity in the light of a benefit—yet, as an abstract question, every one must admit the significance which lies in an unstinted, noble, prodigal abundance. Books, for example : can we have too many of them, provided they be well selected? Dogs: can they be too populous in our grounds? or horses—in our stables? or friends—at convenient distances? or children—in the nursery? or creditors,—no, not creditors, unless gathered together in a general cataclysm. In a word, is not abundance in and for itself a grand advantage? Painfully the truth obtrudes itself upon me as I sit eyeing the solitary Anemone which mopes in a single vase upon my table, the last rose of summer, all its blooming companions dissected and dead. My thoughts take wing to Ilfracombe and Tenby, where foot-pans, pie-dishes, soup-plates, and vases, were crowded with specimens of every variety of form and colour. I think of that paradisaic abundance, and sigh over this one unhappy animal—the mere Mesopotamian pennyworth—partly because I love plenteousness in all things, but mainly because it is only with abundant specimens at command that Nature can be properly interrogated. Fortunately I made good use of my specimens, but not so much as I could make now ; and from my notes I will select a few points for the student's consideration ; but as they will refer solely to questions of anatomy and physiology, the reader is advised to skip the chapter, unless he feel some interest in such questions.

Perhaps nothing has excited more surprise on the part of the public, and nothing has been more unanimously believed by anatomists, than the hypothesis that certain minute organs found in all Polypes, and variously styled *thread-capsules, filiferous-capsules,* or *urticating cells* (Plate III. fig. 5), are organs of urtication, or stinging. The uncritical laxity with which this hypothesis has been accepted may point a lesson. I do not allude to the acceptance of the *fact* that certain capsules containing threads are found in Polypes; but to the acceptance of the alleged purpose, or function, of these capsules. The things are there, sure enough; but whether they serve the urticating purpose, is another matter. Ever since they were first described by Wagner,[*] Erdl,[†] Quatrefages, and Siebold,[‡] they have passed without challenge. They have been detected in the whole group of Polypes, in Jellyfishes, in the papillæ of Eolids, and, according to Van der Hoeven, in Planariæ; yet, as far as my reading extends, not one single experiment has been made to prove the function so unanimously admitted, not a single test has been applied to strengthen or controvert what was, indeed, very plausible, but only *plausible*, not *proven*. Accordingly, no sooner did I submit the question to that rigorous verification which Science imperiously requires, than it became clear to me that my illustrious predecessors—Wagner, Erdl, Siebold, Quatrefages,

---

[*] Wiegmann's *Archiv.*, 1835, ii. p. 215.
[†] Mueller's *Archiv.*, 1841, p. 423.
[‡] *Comp. Anat.*, i. p. 39 (English Trans).

Ehrenberg, Agassiz, and Owen—men whom the most presumptuous would be slow to contradict, had admitted the point without proof, because it wore so plausible an air. Let me hope the reader will accuse me of no immodesty in thus controverting men so eminent; he will see that whereas they have only hypothesis on their side, I have the accumulated and overwhelming weight of experimental evidence.

What are these " capsules," or " urticating cells?" The uninstructed reader may be told that all the Polypes are supposed to urticate, or sting, like nettles; and the nettling organs, or urticating cells, are supposed to be minute suboval microscopic capsules, quite transparent, containing within them threads coiled up, which, on pressure, dart out to many times the length of the capsule, into which they never return. This thread Agassiz likens to a lasso thrown by the Polype to secure its prey. I will not enter here into minute details of structure, which would only confuse the reader, who, if curious, will find all that is known in the works of Mr Gosse, or in the treatises of Owen, Siebold, and Rymer Jones. Any one who has once seen these threads under the microscope darting out with lightning rapidity, especially if he uses a high power, will at once admit that the hypothesis of the " nettling" or "urtication" being performed by these threads is an hypothesis so obvious, an explanation so natural, that —it should be doubted. In all complex matters, we should mistrust the *obvious* explanation; I do not say that we should disregard, or reject it, but mistrust it.

When we know, on the one hand, that the Jellyfish stings, and when, on the other hand, we know that it is furnished with numerous capsules, in which are coiled threads, to be seen darting out when pressed, the idea of connecting the stinging with these threads is inevitable; but this is not enough for Science: it is only a preparatory guess, which *proves* nothing; it may be right, it may be wrong. I believe it is altogether wrong. We have already seen how erroneous was the supposition that Polypes paralysed their victims with a touch, and that poison was secreted by their tentacles; yet for this supposition there was at least the evidence of partial observation, whereas, for the supposition we have now to consider, there is absolutely *no* evidence at all.

On a survey of the places where these "urticating cells" are present, we stumble upon an unlucky fact, and one in itself enough to excite suspicion. They are present in a *few* Jellyfish—which urticate; in Actiniæ —which urticate; and in all Polypes—which, if they do not urticate, are popularly supposed to do so, and at any rate possess some peculiar power of adhesion. In all these cases, organ and function may be said to go together. But the cells are also present in the majority of Jellyfish, which do *not* urticate; in Eolids—which do *not* urticate; and in Planariæ—which do *not* urticate. Here, then, we have the organ, without any corresponding function; "urticating cells," but no urtication! The cautious mind of Owen had already warned us that there was something not quite satisfactory in our supposition ; " some superaddition to the thread-cell would seem to be essential to the urticating faculty," he says, when speaking of the Jellyfish, " since these cells are present in species and parts that do not sting." It is to be regretted that he was not moved by this doubt to a closer examination of the evidence on which the urticating faculty rested ; he would assuredly have been led to the belief that no superaddition to the thread-capsule will account for the phenomenon.

But I waive the argument derived from such a source, and, confining myself to the Anemones, ask the reader what he thinks of this awkward fact, namely, that these urticating capsules are most abundant in parts which do *not* urticate ? Only the tentacles have this power, and although they have numerous capsules, the urtication cannot well be attributed to *them*, since these capsules are more abundant in the " convoluted bands," in the lining wall of the stomach, and in the blue spots which surround the oral disc in the Smooth Anemone—these spots, indeed, being made up of such capsules and small granules—yet in not one of these parts can the slightest urtication be traced ! How is this ? If these capsules are the nettling organs, why do they not nettle in those parts where they are most abundant ? No one has thought of asking this question.

It thus appears that many animals having the capsules, have none of the power attributed to the capsules; and that even in those animals which have the power, it is only present in the tentacles, where the capsules are much less abundant than in parts not manifesting the power: the conclusion, therefore, presses on us that the power does not depend upon these capsules.

And this conclusion is strengthened every step we take. Thus the Anthea is of all Anemones the most powerfully urticating; yet, if we compare its capsules with those of other Anemones, we find them greatly inferior in quantity to those of the Daisy and Dianthus, and much inferior in size to those of Crassicornis, as well as less easily made to uncoil their threads. It has not been remarked, that whereas according to theory, the thread should dart out almost instantaneously on the slightest pressure; in point of fact it frequently cannot be pressed out at all, even when the whole force of the finger is exerted on the two pieces of glass between which it lies. From the very capricious way in which the threads dart out while under the microscope, and not under pressure, and from the frequent impossibility of pressing them out, I suspect that pressure has really nothing normally to do with the ejection of the thread.

Hitherto we have merely considered facts of Observation; we shall now see them confirmed by Experiment. Mr Gosse proposes to establish a new genus, named Sagartia, on this purely hypothetical function; including in it all those Anemones which, like the Daisy and Dianthus, possess an abundance of peculiar white filaments, visible to the naked eye, which are protruded from the pores of the body and the mouth, when the animal is roughly handled. These filaments are seen, on examination, to be chiefly composed of the "urticating cells." Mr Gosse names the genus Sagartia, because Herodotus says of the Sagartians, that "when they engage with the enemy they throw out ropes which have nooses at the end, and whatever any one catches he drags towards himself, and they that are entangled in the coils are put to death." The name, you perceive, is aptly chosen,—that is, it would be, if the hypothesis of the filaments were not a figment. The filaments have no such lasso-like and murderous power. This Mr Gosse would deny; and I remember he somewhere records an observation which would perhaps quite satisfy him that his denial has good ground to stand on. He relates that he once saw a small fish in the convulsions of agony, with one of these filaments in its mouth; it shortly expired. It is a matter of surprise and regret that Mr Gosse, having once made such an observation, did not feel the imperative necessity of repeating and varying it, so as to be sure that the death was not a mere *coincidence*. If the filament had the power which this single observation fairly seemed to suggest, nothing could be easier than to establish the fact by experiment. But, I repeat, no one has seen the necessity for the verification of an hypothesis so plausible; and Mr Gosse, like all his predecessors, was content with recording his observation, as if it carried the point.* Not being so content, I tested

* How little reliance is to be placed on such an observation may be gathered from the following: One evening, at the Scilly Isles, I was startled from my reading by a commotion in the pan on the work-

it thus: After irritating a Dianthus till it sent out a great many filaments, I dropped a very tiny Annelid among them, and entangled it completely in their meshes. Yet lo! these filaments, which are said to possess so powerful a faculty of urtication that even vertebrate animals are killed by them, had no other effect upon a soft Annelid than that of detaining it in their meshes, from which it shortly freed itself, and wriggled away unhurt. Nor was I yet satisfied; placing a tiny Crustacean, of the shrimp family, among the filaments of another Dianthus, I saw it remain there enveloped, but apparently quite comfortable, not in the least so desirous of escaping as one would expect if it were being "nettled" all over; and, when the jar lurched, it swam away.*

I have since repeated this experiment with Entomostraca and Annelids, without once detecting the

---

table, which contained a couple of fish, a Doris, and a few Polypes, but no Sagartia. On investigation, it turned out that the commotion was produced by a fish (*Ophidium*) six or seven inches long, in the convulsions of death. *In its mouth there was a slight strip of Ulva.* As the weed dangled from its mouth, I was forcibly reminded of Mr Gosse's dying fish, with its filament of the Actinia dangling from its mouth. Should I have been justified in attributing to the sea-weed the power Mr Gosse attributed to the filament? Clearly not. In both cases the relation observed was one of *coincidence*, not of *causality*.

* Since the publication of the First Edition, I have learned from Mr Cooper of London, and Mr Broderick of Ilfracombe, that they have observed the tentacles of an Anemone wither up after coming in contact with the body of a Dianthus, and causing it to emit its filaments. I think it right to record such contradictions to my own observations, although I cannot reconcile them, nor pretend to unsay what has been said. The subject is worth a rigorous investigation.

slightest indication of their being more incommoded by the filaments than they would have been by threads of silk. Mr Gosse, indeed, not only maintains that these filaments are weapons of offence, but he actually suggests that the blue spherules which surround the disc of the Mesembryanthemum may "represent the function of these missile filaments" because they are composed of the thread-capsules.

The hypothesis which assigns to the thread-capsules a function of urtication, or prehension, is an hypothesis without a single fact to warrant it, and is contradicted by the various facts I have just adduced. Ehrenberg has very unwarrantably given an ideal figure of a Hydra in the act of seizing its prey, with the hooks of the thread-cell extended; but, as Siebold truly remarks, the animal is *never* seen thus.

Having shown that the parts most abundantly supplied with these "urticating cells" do not urticate, I can now remove the last vestige of doubt by the fact that the capsule itself from the *tentacle* of an Anemone, when seen to eject the thread and touch an animalcule, does *not* kill or disable that animalcule; a fact I witnessed when examining the capsules under the microscope. This not only gives the *coup-de-grace* to the general hypothesis, but even sets aside that suggestion of Professor Owen's respecting the probable superaddition to the "urticating cell" which is to distinguish it from "cells" in those parts destitute of the power; because here we see the capsule taken from an urticating animal does *not* urticate.

L

The foregoing discussion has had a purpose beyond that of rectifying an universal error—the purpose of pointing a lesson in Comparative Anatomy. The greatest living experimental physiologist, Claude Bernard, has recently insisted with emphasis on the importance of recognising "anatomical deduction" to be a fruitful source of error.* He warns us against attempting to deduce a function from mere inspection of the organ, without seeing that organ in operation, and applying to it the test of experiment. As a case of pure deduction, this hypothesis of the "urticating cells" seemed to command, and did command, instantaneous assent; but on submitting it to verification, we find the hypothesis to be an error. To the philosophical mind, therefore, there will have been an interest in the foregoing discussion greater than any interest issuing out of the mere conclusion respecting the thread-capsules.

Among other things, it will illustrate the need there is for rigorous scepticism, and extended observation, on the part of zoological students. So long as we unsuspectingly accept what is repeated in books, without being assured that the statements are made on sufficient evidence, and so long as we have eyes but observe not, zoological progress will necessarily be slow, in spite of the vast number of excellent observers and workers, who *do* accelerate our progress by genuine work. When I insist on the necessity for circumspect doubt, and verified observation, the reader must not understand me as implying that this necessity is not

* *Leçons de Physiologie Expérimentale*, vol. ii. : 1856.

vividly present to the mind of many zoologists, and of every real worker; for in truth, only by such methods can any solid result be reached, and no one even superficially acquainted with the present state of Zoology will be disposed to underrate the importance and extent of that band of distinguished investigators whose researches daily unfold fresh discoveries. Not, therefore, as throwing any shadow of scorn on these men and their methods, nor as if I were bringing a neglected principle into prominence, am I tempted to insist on the only method of successful pursuit in these studies; but simply to distinguish those students of Zoology who wish to increase the circle of knowledge by some small addition of new fact, from students who wish merely to ascertain what is known. In Zoology, as in all other departments of intellectual activity, there are men contented with "information," whose ambition never passes beyond erudition. They want to know what is known. Others there are who, less solicitous, it may be, about what is known, are intensely moved to know for themselves; and these are the workers who extend the circle of the known.

What is known of the reproductive system of Anemones? Not much, and that little confusedly. Our English text-books are somewhat precise; but the precision is for the most part that of error. I carried with me to the coast this amount of definite error, which gradually revealed itself as error in the course of a series of investigations. That the reader may follow clearly the course of reasoning presently to be traced, it is necessary to begin with a few explanations, which the better-instructed will pardon.

Let us first fix in our minds a definite idea of the structure of the Anemone, as far as it will be involved in the subsequent remarks. Imagine a glove expanded into a perfect cylinder by air, the thumb being removed, and the fingers *encircling*, in two or three rows, the summit of the cylinder, while at the base the glove is closed by a flat surface of leather. If now on that disc which lies within the circle of fingers we press down the centre, and so force the elastic leather to *fold inwards*, and form a sort of sac suspended in the cylinder, we have by this means made a mouth and stomach; we then cut a small hole at the bottom of the sac, and thus make a free communication with the general cavity. We then divide this general cavity by numerous partitions of card attached to the wall of the cavity, and form a number of separate chambers called the *interseptal spaces*. Just as the cavity of the finger is continuous with the cavity of the glove, so are the cavities of the tentacles continuous with the interseptal spaces. In these spaces will be found long coils, which are sometimes seen lolling out of the mouth, and always bulge out when the Anemone is cut open; these are called the *convoluted bands*, and to them attention is particularly directed. If the reader will now look at the diagram given in Plate III. fig. 1 (wrong in several details), and also fig. 2, which is an accurate section, and shows a portion of the stomach, the convoluted bands, and under them the ovaries—he will have a tolerably accurate conception of the general structure of an Actinia.

Having given a rough outline of the principal characters of the Actinia's internal structure, sufficient to render intelligible what will hereafter be referred to, I must direct the student anxious for more precise details, to the *Mémoire* by M. Hollard,* as the latest and best anatomical essay; noticing, by the way, that there still remains to be written a *comparative* anatomy of the various Actiniæ, some of which differ in important characters from the others. In the *Anthea*, for example, the tentacles have an abundance of round yellow-brown globules, which make them incapable of being retracted under *normal* conditions; and the same is noticeable in the two horn-like tentacles in *Actinia bellis*. What are these globules? There are several other points of difference to be noted, but I content myself with indicating the desideratum of a careful comparison, to complete our knowledge of the anatomy of this genus. Meanwhile let us endeavour to form some distinct conception of the mode of Reproduction exhibited by the Actiniæ.

Certain general facts must be borne in mind. First, let me call attention to the fact that in all animals, the highest as the lowest, the *envelope* is of eminent importance, its predominance bearing a precise ratio to the simplicity of the organism. The simplest organisms breathe, exhale, secrete, absorb, and reproduce by their envelopes alone; and if the more com-

* *Annales des Sciences*, 1851.

plex organisms perform each of these functions by a *special* apparatus of organs, yet these organs themselves are originally developed *from* the envelope. We may, ideally, reduce even a mammal to a cylindrical envelope folded inwards at each end; from the enfolded skin are developed all nutritive and reproductive organs, while the nervous system and its osseous sheath are developed in the space between the outer and inner walls of the envelope.

"We may, in an ideal manner," says Professor Draper, "conceive the production of the more elementary animal forms, as arising from a simple sac or bag, which, furnishing a starting-point, exhibits its first acquirement of localisation of function, by the doubling of one half into the other, thereby giving rise to a cup or pocket-shape form, so that respiration and digestion, which were confusedly and conjointly carried forward upon the same surface, are now parted from each other, the outside of the cup being devoted to the one, and the inside to the other. Increased endowments are obtained by crimping or dividing the edge of the cup, prehensile organs of less or greater length and power arising thereby; and this in reality is the structure of the Hydra. Another advance is made by the preparation of new and complicated structures, fashioned out in the substance between the inner and the outer wall, and in this manner arise the various mechanisms for respiration and reproduction. Such a state of things is presented by the Actinia."* What

* DRAPER: *Human Physiology*, 1856, p. 501.

is here said respecting Respiration is very questionable, but the general idea of an increasing specialisation of function, in the increasing separation of the parts, is well expressed. It is indeed a fundamental law that every advance in complexity of organisation takes place through a gradual differentiation, or specialisation, of the general envelope. These important synonyms. *differentiation* and *specialisation*, I will explain by illustrating the law to which they point—namely, the law of animal development first enunciated by Goethe, and strikingly applied by Von Baer*:— *Development is always from the General to the Particular, from the Homogeneous to the Heterogeneous, from the Simple to the Complex; and this by a gradual series of differentiations.*

When we say an organ has been formed out of a tissue, we say a differentiation has taken place; and the function, *e. g.* respiration, which before was performed by the general tissue, is now *specialised, i. e.* performed by that special organ. A homogeneous mass of organic matter, such as the Amœba, which has no organ whatever, performs all the functions of Assimilation, Respiration, Locomotion, and Reproduction, by its general mass, not by any special organs.

The process of differentiation by which special organs are gradually developed in the ascending scale of the animal series, is equally exhibited in any particular case of development. Thus if we follow the

---

* GOETHE: *Werke*, xxxvi., *Zur Morphologie*, 1807. VON BAER: *Zur Entwickelungsgeschichte*. 1828. I. 153.

formation of the human hand, we find first a differentiation between the carpus, or wrist, and the metacarpus, or hand; next the fingers are differentiated, but, without any division into separate segments—this takes place later; then we have a separation between the soft and hard parts, the cartilage separating from the plastic mass; then these cartilages become osseous; and in the soft plastic mass we distinguish differentiations into muscle, tendon, skin, &c.; when the single tissues are thus separated we may begin to trace differentiations in the skin, such as the papillæ, the secreting glands, and so forth: till, from a homogeneous mass of cells, we have traced the development of that marvellous and complex structure, the human hand.

Applying this torch to the obscure question of the reproductive system of the Anemones, it at once discloses to us that the Anemone, being of a very simple organisation, we shall be wrong if we expect to find in it a high complexity of special organs. Anatomists, indeed, have often neglected such a consideration, and have worried themselves in the search after organs, which *a priori* were not likely to be present. They have sought for and "discovered" nerves and ganglia, each discoverer scornfully rejecting the alleged discovery of his predecessor, and declaring the nerves were in a totally different locality, while no one anatomist could find them anywhere after another. They have worried themselves about the Respiration of the Anemone, not perceiving that Respiration, like Circu-

DISCOVERY OF THE OVARIES.

lation and other functions elsewhere dependent on a special apparatus, was here performed in a direct and general manner. They have not suspected that Reproduction takes place in the Anemone, much in the same way as in the fresh-water Polype—not in any special and *permanent* apparatus of organs, such as ovary, oviduct, &c., but by a *temporary* specialisation of the general envelope, including an accumulation of germ-cells and sperm-cells. I am aware that special organs called ovaries are described in all books, and that some writers describe an oviduct—which only exists in their imagination, for no duct of any kind is found. Of course, no philosophical *a priori* conclusion could be permitted to stand up in contradiction to observed fact; if the organs are there, it is of no use deductively establishing their non-existence. But *are* they there?

When I first commenced the investigation of Anemones, I had no reason whatever to doubt the statement so generally and confidently made, that the *convoluted bands* were the organs in question.—(Plate III., fig. 3, represents a convoluted band attached to the border of the membrane called the mesentery; the grape-like mass is the ovary.) At the end of the first week my doubts began. These convoluted bands contained no trace of ova, but instead thereof they contained vast quantities of those *thread-capsules* which I then believed to be urticating cells. This was the last place in the world where one might expect to find offensive weapons; and misled by the belief in these cells, I was led to question the function of the convoluted bands. Questioning, of course, meant something more than supine doubt. I began on the 14th May to examine closely into the evidence, and on the 12th June I was fortunate enough to confirm all doubts by the discovery of the real ovaries (such as they are) in a large Crassicornis : here there were no thread-capsules, but abundance of unmistakable ova, each with its " vesicle of Purkinje." The thrill of delight with which the assurance broke upon me may be conceived.* At that time I, of course, believed that the grape-like cluster in which the ova were lying were true and permanent ovaries ; but having since been frequently unable to detect them in adult specimens, and never in young specimens, I came to the conclusion that these ovaries are *temporary* organs, formed by an accumulation of germ-cells in various parts of the free border of the septa; that, in fact, they represent the first rudimentary state of what in higher animals becomes the special organ. This conclusion is, however, purely theoretical, and I will now state what any one may see who examines an adult fresh from the rock-pool or tank. With a rapid but not deep incision we lay open the envelope from the outside; the convoluted

---

* I subsequently ascertained that Mr Teale and M. Hollard had described these ovaries ; I only claim priority in the elucidation of their structure and function, as temporary organs, similar to those of the Hydra. [Since this was published, it has been found that the sexual organs of the smaller Naïds are also only temporary, and disappear after the ova and spermatozoa are extended.—See GEGENBAUR: *Vergleichende Anatomie*, 1859, p. 183.]

bands will bulge through the opening; but if we are vigilant and brush these aside, we shall perceive certain lobular or grape-like masses of darker colour, almost entirely hidden by these bands, but growing from the septa.—(Plate III., fig. 2, represents a section of the Actinia, showing the ovaries lying under the convoluted bands, and attached to the septum; fig. 3, the ovary, when spread out on a glass slide.) They are not situated in any precise spot; near the base, about the centre, and close to the disc, they may be found: nor are they on every septum; sometimes we may make three or four incisions before detecting them.

Such are the ovaries of the *Crassicornis*: but are they entitled to the name? Are they organs at all? A minute inspection of them will confirm what I said just now that they are *not* " organs," properly so called; that is to say, they are not, like the ovaries of higher animals, permanent organs having a definite and *specific* structure; they are, in truth, nothing but accumulations of germ-cells in a delicate membrane. They have none of the essential characteristics of an ovary. It is true that Spix, Delle Chiaje, Rapp, de Blainville, Van der Hoeven, and others, describe what they call oviducts, without, however, agreeing as to their disposition. But Mr Teale and M. Hollard have been unable to find them, and I also can confidently assert that no duct whatever, nor anything distantly resembling it, exists; but, as I have convinced myself by scores of dissections, the whole structure of the ovary is limited to a delicate membranous stroma, in which the ova are imbedded. When the ova are matured, it is most probable that this stroma bursts to set them free, and they fall into the general cavity, where their further development takes place. It would be justly considered an unwarrantable laxity in scientific language if the temporary accumulation of germ-cells beneath the investing membrane of the Hydra were designated as an ovary; and no less unwarrantable is it to call a somewhat similar accumulation of germ-cells in the Actinia, an ovary. The pretended "organ" is not permanent, it is not even constant in its locality, for it may be found at any part of the free border of the septum; and, finally, it has no specific structure, which distinguishes it from any other part of the lining membrane; its grape-like form is owing entirely to the ova imbedded in it.

The reader perceives that I regard the organisation of the Actinia as much simpler than other writers seem willing to admit, and that, in consequence, I interpret its functions more in accordance with the laws which regulate the simpler organisms. He will hereafter see that such a point of view led me to examine and disprove the current notions respecting the Digestion and Nutrition of the Actiniæ,[*] and he will now see that it led me to the discovery of their true sexual character.

The universal belief is that the Actiniæ are of separate sexes, and Kölliker, to whom most subsequent writers refer, asserts that the males are about as numerous as the females. The great difficulties in the way

[*] Part. III., Chaps. I. and II.

of observation can only be estimated by those who have tried it. Rapp declared there was no trace of the male organs. "All the Actiniæ are females," he says boldly. Wagner having discovered, as he thought, spermatozoa in the convoluted bands, pronounced those bands to be the reproductive organs of the male; and this statement was accepted and repeated from book to book. It is now known that Wagner was mistaken;[*] perhaps he had removed a portion of the ovary with the convoluted band, and in that case he might have seen true spermatozoa. "What!" exclaims the reader, "spermatozoa in the ovary? And pray, sir, how *gat* they there?" They were *generated* there, is my simple answer; but before such an answer will be received a few remarks are necessary.

Believing in the simplicity of this animal's organisation, and disbelieving in the existence of any proper ovary, I was naturally led *a priori* to disbelieve in the existence of male reproductive organs. If a temporary specialisation of the lining membrane sufficed as an ovary, would not a similar specialisation suffice as a testis? Meditating on this point, it occurred to me as very probable that the same spot of the lining membrane might subserve both purposes. What said Fact to so plausible a Theory? At first it answered by what seemed like blank negatives; on three fortunate occasions, however, it seemed to answer in unequivocal affirmation, for I found real spermatozoa moving amid

[*] This leaves the function of the convoluted bands a mystery. In Part III. Chap. II., towards the close, the question is touched on.

the ova on the stage of the Microscope. The observations were neither sufficiently extensive nor sufficiently removed from opposite interpretations to admit of a very positive statement on the point; accordingly I contented myself with indicating my belief* that ova and spermatozoa were intermingled in the same stroma, announcing my intention of more attentively investigating the point when next at the coast. Meanwhile, I had read the *Mémoire* on the *Cerianthus*, an animal nearly allied to our Anemones, published by M. Jules Haime,† in which there is a detailed description, with diagrams, of this very disposition of ova and spermatozoa in the same "organ;" so that if it had not been for the positive statement of Kölliker, respecting the separation of the sexes, I should have conceived the point placed beyond dispute.

At the Scilly Isles and Jersey my investigations were renewed, for a long time with the discouraging and paradoxical result of finding ova and nothing else. It seemed as if Rapp's statement would turn out to be correct, and only females were to be found. But then, whence the spermatozoa seen at Tenby? Whence those described by Kölliker? That the absence of spermatozoa did not, in the simpler organisms, by any means imply the absence of reproduction, but that females were capable of propagating, unaided, I knew well enough.‡ This did not answer the question raised. The

---

* *Blackwood's Magazine*, Jan. 1857.
† *Annales des Sciences*, 1854.
‡ See examples in Part IV., Chap. I.

fact that there were sometimes spermatozoa seemed beyond dispute; Kölliker had seen them, and asserted them to be as numerous as ova. The solution of the difficulty dawned upon me when I discovered that the spermatozoa were contained in vesicles, so like ova in the early stage as to be easily mistaken for them. It is only by crushing these that the spermatozoa are seen escaping. Thus it became intelligible that I might have had spermatic vesicles in the ovaries under observation, where I thought there were only ova; and all my females *might* have been hermaphrodite. Unhappily this clue was detected at a period when I was so deeply engrossed with the nervous system of the Mollusca, that I had little time to devote to the unravelling of the whole mystery. But I made a few direct and conclusive observations, which assured me that ova and spermatozoa were intermingled in the same stroma; and the student may easily verify these observations when informed that the ova are distinguishable from the spermatic vesicles by their darker colour; he has only to spread out, on a glass slide, a fold of the membrane containing the ova, and he will perceive among them polyhedral and oval bodies of a paler colour: if these are crushed, the spermatozoa will be seen escaping from them.

How can this be reconciled with Kölliker's and Hollard's statement of the sexes being separate? If I could have devoted the requisite time to the point, when specimens were abundant, I might, perhaps, have answered positively what I can now only answer hypothetically. As it is, I propound a suggestion, which the researches of some more fortunate inquirer may confirm, or rectify. First, be it remembered that the ova and spermatozoa are developed in precisely the same portion of the lining membrane on the free border of the septum. This is the statement of Kölliker and Hollard, who further remark that it is *only* by the difference of colour that the testis is distinguished from the ovary. If these anatomists found ova in one animal, and spermatozoa in another, whereas I found *both* in one and the same animal, four possible explanations suggest themselves :

1st. There are Actiniæ in whom the ova and spermatozoa are intermingled ; that is to say, one grape-like cluster will contain both kinds of cells.

2d. There may be others in whom the spermatozoa accumulate on *one part* of the septum, and the ova on *another part* of the same septum ; or different septa may bear different kinds of cells.

3d. There may be others—and this has its parallel in other Polypes, in Crustacea, and in Insects—which produce *nothing but ova ;* and these ova may develop into embryos, without the concourse of spermatozoa at all.

4th. Or finally, the slight indication of the separation of sex, which is presented when one side of the septum, or one of the septa, develops only one kind of cell, may be further carried out in some species, and in *them* the complete separation of sexes takes place.

Of these four possible forms, I have positive observation only of the first, and am strongly inclined to believe in the third; there is, however, nothing at all improbable in the second and fourth, and either, or both, may be ascertained by careful investigation, which would rectify or confirm the observations of Kölliker and Hollard. But it must not be forgotten that, although a single case of distinct separation—a single Actinia found bearing spermatozoa only on *all* its septa—would suffice to establish the truth of the fourth proposition; yet for this to be established, the inquirer must be rigorously certain that *all* the septa bear the spermatozoa, the ova being everywhere absent; and even then no amount of such evidence will invalidate the fact announced in the first proposition, that in some cases the ova and spermatozoa are mingled; a fact I have quite recently discovered also in the fresh-water Polype. Whether Actiniæ are, or are not, in general, of separate sexes—and I think they are never or very rarely so—the fact remains that they are *also* hermaphrodite. Nor can this surprise us now we know that even fish, which are almost universally of separate sexes, also present hermaphroditism as a normal phenomenon in the Perch genus.*

To sum up the various points we have just been considering, it appears that the Actiniæ are of very simple organisation; that they have no sexual organs

---

* See Dufossé, in *Annales des Sciences*, 1857. Compare also Huxley, *On a hermaphrodite and fissiparous species of Tubicolar Annelid*, in the *Edinburgh New Philosophical Journal*: January 1855, pp. 11-12.

at all; and that the temporary organ produced by an accumulation of cells in a part of the lining membrane, contains both the male and female elements, although possibly some indication of a separation of sexes does occasionally present itself.

A proof of the great simplicity of their organisation, and an argument against the separation of their sexes, is seen in the fact of their being able to reproduce themselves from a mere fragment. In a letter received from Mr R. Q. Couch, of Penzance, there is the following passage: "It is said by some that the sexes in these creatures are separate, but I rather think this will be found not to be the case. Speaking of reproduction, I may mention that the small specimen of *Actinia dianthus* which you saw is thriving (Mr Couch refers to a tiny Actinia which had grown from a shred of the base which adhered to a stone when the *Dianthus* was torn away), and that the day you left I made four small cuttings from the bases of the red and white varieties, which, at the present moment, are adorned with two rows of tentacles." And the following report of a communication made by Dr Strethill Wright to the Royal Physical Society of Edinburgh, which is taken from the *Edinburgh Philosophical Journal*, will be read with great interest:—

"The author stated that *Actinia dianthus*, the Plumose Sea-Anemone of Dalyell, was found on the shores of the Firth of Forth, generally on rocks which were uncovered by the sea only at very low tides. Its habitat was not extensive; it is gregarious, great numbers being frequently found in a very limited space. At Arran he had seen several hundreds closely aggregated together, clothing the roof of a wide low cave, and hanging down like so many membranous bags half filled with water. A similar colony had existed on the perpendicular surface of a single large stone opposite to the Baths at Seafield; and again, another on the under surface of a large overhanging rock at Wardie. It had been a matter of question with the author, how the young of these Actinias, if ejected from the mouth, as in *Actinia mesembryanthemum, troglodytes, bellis*, and *gemmacea*, were able to attach themselves to the rocks, instead of falling down and being washed away by the tide. It was known that *Actinia mesembryanthemum, troglodytes,* and *bellis* were exceedingly prolific, Sir John Dalyell and Dr Cobbold having seen twenty or thirty produced at a single litter from the first species, and yet the number of very young Actinias found in situations where old specimens abounded was very small, and certainly bore no proportion to the number generated. The cave at Arran was very difficult of access, on account of its shallowness and the floor being covered by a pool of water; and the Actinias were only to be reached by assuming a posture which could not be maintained for more than a few minutes. A number were, however, obtained, which, being attached to sponges, were easily stripped from the rock, and with them were associated a great number of very small specimens. Not long afterwards the author noticed a number of young surrounding a large white *dianthus* in the Vivarium of a friend at Leith, and was told that the Actinia, while moving round the tank, had left behind it small white bodies, which separated themselves from the foot or sucker and became young Actinias. Sir John Dalyell had described a similar mode of multiplication in *Actinia lacerata*, and Hollard in *Actinia rosea* (?) The former writer had observed that *Actinia lacerata* protruded from all parts of its foot, stolons or suckers, which became detached, and presently put forth tentacles, and were developed into minute Actinias. After reading Sir John Dalyell's account of *Actinia lacerata*, Dr Wright was anxious to ascertain whether there might not be included in the prolongations separated from the foot, either true ova or germs, or some tissue specialised for the production of young. In the hydroid zoophytes, such as *Hydra*, *Coryne*, &c., the walls of the body consisted of three elements or layers,—a dermal or integumental, an areolar or muscular, and a mucous or intestinal layer; and when gemmation took place in these animals, it occurred by the protrusion of a simple diverticulum or sac from the canal of the body, formed of all the three elements. This diverticulum was developed into a polype body, with mouth and tentacles like those of the polype, from which it pullulated; the two bodies having the digestive canal and all the tissues continuous with each other. In *Hydra tuba* multiplication took place by stolons, which extended to some distance from the body before the new polype bodies sprouted from them; but in that case also a prolongation of the intestinal element passed through the stolon from the old into the new body. These new polypes were not young; their production was a simple increase of the individual, becoming afterwards a multiplication, either by accident, in some cases, or in others by a natural process of absorption. The structure of the helianthoid zoophytes or Actinias was more complicated in its development than that of the hydroid polypi, but it consists of the same three elements, The dermal coat was succeeded by the muscular element, which constituted the chief part of the external wall of the body and tentacles, and then passed inward to the stomach, in the form of septa or partitions, which suspended that viscus in the centre of the body, and divided the intervening spaces into numerous chambers. The mucous or intestinal element existed as a flattened sac or stomach, which appeared, when viewed edgeways, as a mere line extending down about half the centre of the body. The stomach communicated freely with the general cavity of the body. This cavity, which corresponded to the water-vascular system of the Acalephæ, was single below, but as it passed upward it formed a number of chambers divided from each other by the septa before mentioned, and finally communicated with the tentacles, each chamber terminating in the cavity of a single tentacle.

"The whole of the general cavity and its chambers was lined with cilia, by which a constant circulation of the fluid was sustained, and the functions of nutrition, respiration, and excretion were all carried on simultaneously. From the lining membrane of the general cavity, the male and female reproductive organs were also developed, and there, in some species, the ova were hatched, and the young (at first mere shapeless, ciliated germs, swimming rapidly in the fluids of the cavity, chambers, and tentacles) became fully formed, passed into the stomach of the parent, and were ejected from the mouth as perfect Actinias, with mouth, tentacles, and suctorial foot. The author had thought it possible that the prolongations from the foot of *Actinia lacerata* might contain one of these hatched germs in its imperfect state, and that it might be thus deposited on the surface occupied by the parent, and its safety insured. Having some specimens of dianthus in his possession, he had waited for some time in vain for their multiplication by fissure; he therefore determined to try an *experimentum crucis*, and for that purpose having placed the specimen in a jar of sea water, and fed it until it had become fully distended, he examined the edge of the foot, which was perfectly transparent, with a powerful lens, and convinced himself that no ovum or germ existed in that situation. He then separated a piece about a line in length, by half a line in breadth, from the edge of the foot. The parts immediately receded from each other, and the next day he found that the separated portion had crept to a considerable distance along the glass. In two or three days it had raised its divided edge from the surface to which it was attached, and had become a curved column; in a fortnight tentacles had appeared; and in three weeks it had become a perfect Actinia, with a single row of beautiful long tentacles. From the foot of this small Actinia he cut two other exceedingly minute slips, which also became Actinias; and from the foot of the original Actinia he also separated, at various times, fourteen other slips, all of which became developed as the first. The author stated that this case of gemmiparous increase was an instance of the development of a perfect and very complicated organism, from a minute fragment of one similar to itself, all that was essential to the process being apparently the existence of a portion of each of the three elemental tissues of the original, the dermal, the muscular, and the mucous tissue, —the last being represented by the lining membrane of the general cavity. And it appeared to be analogous to the instance of gemmation from the water-vascular system observed by the late Professor Edward Forbes, in *Sarsia prolifera*, in which animal the young medusæ pullulated forth from the hollow bulbs which supported the tentacles."

It should be borne in mind, however, that the Actiniæ are far less capable of reproduction from fragments than the fresh-water Polypes are. Delle Chiaje, indeed, emphatically denies that they can do more than reproduce their tentacles.* This, as we have just seen, is not the case; and it is somewhat curious that Delle Chiaje should confine the reproductive power to the tentacles, when we remember that, according to Owen,

---

* DELLE CHIAJE: *Descrizione e notomia degli animali invertcb. della Sicilia*, iv. p. 130.

the tentacles are the only portions of the fresh-water Polype which are incapable of reproducing the whole organism when separated from it.*

But we must bid adieu to the Anemones for the present, attractive as they are. These pages will have shown that much yet remains to be done before a clear and positive account of their structure and functions can be written; and as they are just now "the rage," we may hope that abundant experience will issue in some satisfactory result.

* I am informed by Mr Bohn, of Essex Street, that, in the tank of one of his customers, an Anthea spontaneously divided itself into two—"both mother and child doing as well as can be expected." Mr Beck, the optician, also informs me that he has watched the process of self-division in the Anthea, beginning at the base, and ending in the production of two perfect animals.

PART III.

THE SCILLY ISLES.

CHAPTER I.

THE LION THAT HAS EATEN A MAN, AND THE ZOOLOGIST WHO HAS BEEN AT THE COAST—TROUBLESOME DESIRES—CHOICE OF THE SCILLY ISLES—PENZANCE LODGINGS—THE SAIL TO SCILLY: PURSUIT OF KNOWLEDGE UNDER DIFFICULTIES—FIRST SIGHT OF THE ISLANDS—THEIR AREA AND POPULATION—THEIR PICTURESQUENESS—THE CHANGEABLENESS OF ROCKS—ANTIQUITIES OF SCILLY—THE INHABITANTS—PRIMITIVE STATE OF THE COMMERCE—DINNER DIFFICULTIES—HOW THE TEN THOUSAND SALUTED THE SEA—LOVE OF THE ENGLISH FOR THE SEA—HOMER—OUR FIRST DAY ON THE ROCKS—THE NYMPHON GRACILE—THE COMATULA—ON OBSERVATION AND EXPERIMENT IN BIOLOGY—DO THE ANEMONES DIGEST?—MEANING OF DIGESTION—ASSIMILATION AND DIGESTION—THE ACTINOPHRYS—FOOD AND BLOOD—EXPERIMENTS ON THE ANEMONES—FOOD AND KNOWLEDGE.

BETWEEN the lion that has once eaten a man—once tasted the glory and ambrosial delight of man-beef—and the lion remotely ignorant of that flavour, there lies a chasm. Only in zoological text-books can the two animals be considered as of the same species. In profounder characteristics, in the complexion of their souls, they differ as the Caucasian differs from the Hottentot. The lion who has once fed on man, carries with him an unforgettable experience; he has supped with the gods, and Homeric rhythms murmur in his ears. Visions of that ecstatic hour hover before him in his lair, accompany his moonlight marches through the mountain-gorge, thrill him with retrospective flavours as he laps the moonlit lake, and fill with a certain blissful torment all his leisure moments. These visions, like the after-glow of sunset on the Alps, tinge his mental horizon, and create a gustatory after-glow which warms his whole frame. Haunted by such recollections, tormented by the appetites they develop, his nature undergoes mysterious, modifying influences; new and grander ferocities are awakened, which, in turn, develop fiercer daring, and render him ten times more formidable. Hitherto he has wanted something of the daring commensurate with his strength. He has always avoided personal combat with an European, when honourably the challenge could be ignored. But now the case is very different; now, the scent of human blood thrills along every fibre; and when sight reveals the proximity of his noble foe, then flashes the tawny eye with sombre fire, the terrible talons tear up the earth, he dresses his mighty mane, and prepares for the fight in slow, solemn, concentrated wrath, clearly foreseeing that two issues, and only two, remain open for him—man-beef, or a tomb.

Not less profound, although not quite so terrible to his enemies, is the difference between the man who has once tasted of a noble sea-side passion, once lived with his microscope for a few months on the wealthy shores of some secluded spot, indulging in a new pursuit—

and the common man, utterly remote from all such experience, walled out from it by blank negation, incapable of even conceiving the heights and depths of such a passion. Visions of those ecstatic hours for ever accompany the happy man. He may return to his home, and resume the labours of his profession, which secures him pudding, and, it may be, praise: he continues the daily round, but not as before. He is a changed man. The direction of his thoughts is constantly seawards. Murmurs of old ocean linger in his soul, as they murmur in a shell long since taken from the deep, and now condemned to ornament the mantel-piece of some lodging-house, the daily witness of prosaisms and peculations. To the casual eye he may not seem changed; but read his soul, and you will find he is another man.

At least it was thus with me. I had supped with the gods, and grew fastidious over my shilling ordinary. If work imperiously claimed my attention, if I was forced to trouble myself with "proofs," commentators, old writers, dreary philosophies, and multiform affairs, the glass vases on my table, perpetually reminding me of Ilfracombe and Tenby, aggravated the oppression. The iodine of the sea-breezes had entered me. I felt that I had "suffered a sea change" into something zoological and strange. Men began to appear like molluscs; and their ways the ways of creatures in a larger rock-pool. When forced to endure the conversation of some "friend of the family," with well-waxed whiskers and imperturbable shirt-front, I caught myself speculating as to what sort of figure *he* would make in the vivarium—*not* always to his glorification. In a word, it was painfully evident that London wearied me, and that I was troubled in my mind. I had tasted of a new delight; and the hungry soul of man leaps on a new passion to master, or be mastered by it.

"Chacun veut en sagesse ériger sa folie"

says Boileau, and I was willing enough to demonstrate to all recusants that my passion had a most rational basis. Meanwhile it was the torment of intellectual hunger; and I make it a rule always to satisfy hunger —on philosophical principles. If you don't content *it*, it will torment *you;* it obtrudes on work and duty, perplexing the one, and obstructing the other: it can't be starved into silence. When pastry-cooks hire new boys, they wisely permit an unrestricted glut of tarts. The young gluttons fall on, tooth and nail, and in a week are surfeited; whereas a stealthy and restricted appetite would have lasted them for years, much to the damage of the pastry-cook. In this philosophic forethought I resolved to give myself a glut of zoology, to let loose the reins of desire, and afterwards, if the fates so willed it, settle once more into a student of books, and writer thereof. It was really time. For seven long months I had been separated from the coast; and like the Cyclops of Euripides, I had grown weary of feeding on daily butcher's meat and game, just like other stray mortals in the Strand; and smacked my lips at the prospect of man-beef. With the Cyclops I exclaimed :—

> "I am quite sick of the wild mountain-game;
> Of stags and lions I have gorged enough,
> And I grow hungry for the flesh of men."*

March was already come, the equinoctial gales were near, and the Isles of Scilly beckoned like syrens from their dangerous shores. The weather was intensely unlike summer, the snow and hail freely falling; so that, on a first blush, there did seem a shadow of reason for the astonishment of friends, who looked upon departure at such a time, and for such a place, as indicating something like insanity. But great wits to madness nearly are allied, and this alliance with great wits will perhaps be granted to me. At any rate there was method in the madness, for unless I reached the coast before the equinox, the passage would be more than usually perilous; and just after the equinox, as everybody knows, the spring-tides recede to greater depths, and offer the finest opportunities for rock-hunting; moreover, the gales at this period throw welcome treasures on the beach. The 15th of March, therefore, was the very latest date I could afford for departure; and on that day the journey began.

Why the Isles of Scilly were obstinately selected,

---

* This is Shelley's translation. The reader who has not quite forgotten his Greek may like to have the original :—

'Ὡς ἐκπλεώς γε δαιτός εἴμ' ὀρεσκόου·
Ἄλις λεόντων ἐστὶ μοι θοινωμένῳ
Ἐλάφων τε, χρόνιος δ' εἴμ ἀπ ἀνθρώπων βορας.

may not be so easily explained. I had a fixed idea on this point; no argument could make me swerve from it. The main attraction was doubtless lurking in my profound geographical ignorance, which invested these Isles with a mysterious halo. In days when ladies take pleasure-trips to Algiers, and reach it in four days, or run up the Nile, as formerly they scampered through France, any real bit of untravelled country necessarily creates an interest; and for travellers, in the adventurous or pleasure-hunting sense, Scilly is as virgin ground as Timbuctoo. Vessels in abundance touch there; but who *goes* there? Indeed, on entering a shop to make a small purchase, the bland woman compassionately inquired whether I had been "driven by contrary winds" to this unfrequented spot; evidently never conceiving the possibility of a sane Englishman *coming* here. They are also difficult of access: "a very dangerous flat, and fatal, where the carcasses of many a tall ship lie buried." Ten days, owing to contrary winds, were consumed in getting here; and under the most favourable conjuncture of trains, coaches, and winds, three days would be the very shortest time required. This difficulty secures the place from the nuisance of "visitors." Moreover, I had an idea of its being a good spot for zoological research; and with these two advantages, I could afford to listen unmoved to the sarcastic questions pelted at me, such as: Can you get anything to eat there? Are the Islands inhabited? Do the people speak English? Are they civilised?

Contrary winds, and what sailors call "dirty weather," detained me a week at Penzance, where I was stranded in a lodging-house, kept by a middle-aged Harpy, rearing a brood of young Harpies, and rendered all the more fierce in lodging-house instincts by her condition of widowhood, which, you may have observed, generally throws a woman on the naked ferocities of her nature. Were you ever in nautical lodgings? Do you remember their ornaments,—the cases of stuffed birds and fish, the shells on the mantelpiece, and the engravings irradiating the walls: a "Sailor's Departure," with whimpering wife and sentimental offspring; a "Sailor's Return," with joyous wife and capering juveniles? All these adorned my rooms, which were further adorned by a correct misrepresentation of the brig Triton, as she appeared entering an impossible harbour of Marseilles, flanked by a portrait of the defunct husband, master of the aforesaid brig, painted in the well-known style: a resplendent shirt-front with a head attached, sternly inexpressive, on a mahogany background. The defunct mariner seemed blank with astonishment at my courage in coming to such a house—a ruin, not a lodging. Everything in it was afflicted with the rickets. The chair-backs creaked in harmonious threats, if you incautiously leaned against them. The fire-irons fell continually from their unstable rests. The bed-pole tumbled at my feet when I attempted to draw the curtain. The doors wouldn't shut. Even the teapot had a *wobbly* top, which resisted all closing. Nay—and this will surprise you—in the moral world I noticed a similar dilapidation. The discrepancies were painful. In the "bill," arrangements were made which showed fiscal genius: and when a suggestion was offered that the remains of yesterday's fowl might serve for to-day's luncheon, a look of pained reproach passed over the widow's face, followed by a gulp, and a silence which was broken only by diversion of the dialogue into quite other directions—the look, the gulp, the silence expressed, as plainly as words, the mean opinion which the widow entertained of her victim. Low as her opinion had placed him before, it had not reached such depths as *that;* the request for a paltry remnant of fowl, indeed, was answerable only by profound silence. Thus it *was* answered. I never gazed upon that bird again.

Weather-bound in such a place—the equinoctial gales hurrying on—boxes corded, soul unquiet—you may imagine the alacrity with which I sprang out of bed, the morning when a sailor came up from the packet to say that anchor was weighed, and we should start as soon as I could slip on my things. This was at six in the morning, and, by half-past, the Ariadne, formerly Lord Godolphin's yacht, but now the property of Captain Tregarthen, who runs it between Scilly and Penzance as the mail and sole communication, left the harbour, and reached Scilly by one o'clock. This was on Thursday, 26th March 1857. A century ago, on the 25th May 1752, Borlase, the admirable antiquarian, whose *Observations on the Ancient and Present State of the Scilly Islands* was among my books,* set sail in the sloop Godolphin at seven in the morning, and about nine in the evening drew near the islands—drew near, but dared not venture nearer; because, a "very thick fog ensuing, the sailors began to be apprehensive whether they should fall in with the proper passage into St Mary's Island or not: sometimes they thought they could see the land, but were always uncertain what part of the island it was. This determined us to continue turning off and on (in sea affairs give me leave to use sea expressions), and wait for the morning. During this interval we had a very uneasy time of it, and nothing to do but to expect the daylight, which, you may be assured, was with great impatience. The day came, but the fog continued so thick that we had no benefit from it." In this fog they continued beating about, in terror of getting entangled among the narrow guts; but about six the fog cleared, and revealed to them St Mary's Island close at hand. "We were such true sailors," he says, "that we immediately lost sight of the danger we had escaped, delighted as we were with the thoughts of being soon in port, and the uncommon appearance of the land (if what is mostly rocks can be called so) on each side of us as we passed. It was Crow's Sound; and I must own the sight of it gave me much pleasure, which you will, and justly may in some measure, attribute to our sudden transition from

---

* Thanks to that most convenient, and to all students most valuable of institutions, *The London Library*, which manifold experience causes me to urge every man of letters to join.

a state of uncertainty to that of safety, but not wholly; for these islets and rocks edge this Sound in an extremely pretty and very different manner from anything I had seen before. The sides of these little islands continue their greenness to the brim of the water, where they are either surrounded by rocks of different shapes, which start up here and there as you advance, like so many enchanted castles, or by a verge of sand of the brightest colour."

If this was the passage made during gentle May, surely we were very fortunate, in blustering March, to have got over all our troubles in six hours. Shorter, our passage undeniably was; whether it was also sweeter, remains a problem, towards the solution of which I will say thus much,—that under no extension of euphuism could it be called *sweet*. In the first place, there had been no breakfast to begin the day; and the Ariadne offered nothing in the culinary department. Cheering Souchong, or aromatic Mocha, to warm the matutinal ventrals, was not to be thought of; we were even lucky to have a dry biscuit to munch in philosophic resignation. Deprived thus of our natural fortifications against the advancing enemy, we were further disabled by the rain, which forced us to descend into the cabin, and get into our berths. In these exiguous spaces we remained until the joyful tidings of arrival flooded us with sudden energy, and flung the past hours from us like a hideous dream. Except during the brief intervals of sleep and semi-delirium, those hours were not pleasant. The cold, not to be kept out by any amount of rugs, cloaks, and tarpaulin, seemed stealthily creeping into the very centres of life. The sensations which fly around sea-sickness need scarcely be alluded to. Constantly, when my intellect was sufficiently disentangled from these sensations to exercise itself, the thought would arise that pleasanter far was the pursuit of zoology in comfortable homes (where Mr Lloyd of Portland Road, or Mr Bohn of Essex Street, would supply tanks and vases with the desired animals in exchange for vulgar dross, thus bringing the forces of commerce and civilisation to minister to our pursuits), compared with this harum-scarum method of trusting oneself to "sea-traversing ships," in order to become one's own purveyor. This thought would occur. And then the fluctuating intellect passed into self-condemnation at thoughts so base, remorse so ill-timed, cowardice so unzoological. These passing pangs, however unattractive, would they not inevitably pass? And how the released spirit, in its reinstated vigour, would rejoice at having undergone such torments for such weeks of enjoyment.

As I said, the joyful tidings came at last. With alacrity I urged my staggering steps up the ladder, and emerged upon the deck, where the bright sunlight revealed a scene, which of itself was repayment and full discharge for any arrears of misery. We were in St Mary's Sound. The islands lay around us, ten times bigger than imagination had prefigured, and incomparably more beautiful. On their picturesque varieties I might turn a green countenance and glazed eye, but the heart within me bounded like a leopard on his prey. This was worth coming to! Those poor devils who sit at home at ease, and supply their tanks from commercial sources, were now the objects of pitiless sarcasms for their want of enterprise. In such a mood I hastily secured comfortable lodgings, clean as a Dutchman's, at the Post-office; swallowed some tea and toast, to appease the baser appetites, and hurried forth to satisfy the hunger of the soul, by a survey of the Bay, and its promises. The promontory on which stands Star Castle, offered a fine breezy walk over downs resplendent with golden furze,* and suffered the eye to take the widest sweep. How thoroughly I enjoyed that walk! The downs were so brilliant that one could sympathise with the enthusiasm of Linnaeus on his arrival in England, and his first sight of furze, as he flung himself on his knees, and thanked God for having made anything so beautiful. The downs were all aflame with their golden light. Ever and anon a rabbit started across the path, or the timid deer were seen emerging from the clumps of golden bush. A glance at the many reefs and creeks along the wavy shores raised expectation tiptoe, forcing hope into certainty of treasures abounding. Whatever drawbacks Scilly might possibly have in store, this at least was indubitable — the hunting would be good. Not that any shadow of a drawback

---

* The reader who has not seen the furze in Devonshire and Cornwall can form but a faint idea of its rich colour and profusion.

darkened the horizon; for what could the heart desire more? Here was a little archipelago, such as Greek heroes might have lived in—bold, rugged, picturesque,—secure from all the assaults of idle watering-place frequenters,—lovely to the eye, full of promise to the mind—health in every breeze. Ithaca was visibly opposite. Homer's cadences were sweetly audible. Here one might write epics finer than the Odyssey, had one but genius packed up in one's carpet-bag; and if the genius had been forgotten, left behind (by some strange oversight), at any rate there was the microscope and scalpel, with which one might follow in the tracks of the "stout Stagyrite," whom the world is now beginning to recognise among the greatest of its naturalists. Homer or Aristotle? The modest choice lay there; and as Montaigne says—"nous allons par là quester une friande gloire à piper le sot monde." (The *sot monde* being you, beloved reader.)

It is puzzling to determine the number of the Scilly Isles, because, where the largest, St Mary's, is on a scale of no greater magnitude than nine miles in circumference, it becomes a nice point to settle how *small* a patch of rock is to be reckoned as an island. There are some hundred, or hundred and twenty, distinct islets; but of inhabited islands only six. The area in statute acres is 3560, and the population in 1851 was, according to the census, 2600 in 511 houses —the females predominating in the ratio of 1439 to 1162. The average of death is 16 in 1000; in other parts of England it is 23 in 1000, showing a decided hygienic superiority in favour of Scilly. Much arable land there is not, but an occasional upland smiles prosperity at you; and in the sheltered nooks of Holy Vale you are startled with the appearance of what almost looks like a tree. In the other parts of the island no tree is discoverable—without a lens. The lanes are formed of stone hedges, as in Devonshire and Cornwall; but these hedges are not, at this early season, prodigal of ferns and wild-flowers as they will be soon. Yet they have already abundant ornament. On the summit grows the furze, with its profuse bunches of gold; from the crevices peep the stonecrop, the leaves of the fox-glove, pennywort, and a multitude of other wall-loving plants, dear to my eye, though unknown by name; already the dog-violet and celandine are gay with colour, and the lichens tint the stone with delicate pale greys or greens, deep orange, or bright gold.

The grouping of the islands is very picturesque, forming several good Sounds, where vessels of great tonnage find secure anchorage, and give a pleasant aspect to the scene. Standing on any of the eminences, we gaze down upon the deep blue of the bays, the white sweep of sands, and rugged reefs, and purple masses of the opposite shores; the plaint of the seagull, floating overhead, being almost the only sound audible, except the never-ending symphony of the waters. As we ramble round the coast, the successive scenes of the unfolding panorama make us long to have the artist's power of transferring them to our sketch-book. The rocks are entirely of granite; and the huge wave-worn boulders, sudden pillars, and piles of broad ledges into which they have been disrupted, give endless variety to their forms. Sometimes they have a castellated aspect, as at "Giant's Castle," on the southern coast—a noble edifice of nature's cunning architecture. Beautiful are the outlines of its topmost grey shelving ledges, softened with shaggy pale-green Byssus-lichen,—beautiful its huge rectangular masses of warm light-brown, blackened here and there with the mysterious beginnings of life, and darkening downwards to the shining deep-brown reefs that jut from the Atlantic waves, which lift their curling masses of crystal greenness into momentary splendour, and then dash, and break, and whirl in milky eddies among the ever-passive rocks. Passive are they? Yes; and yet passivity itself is only a slower action, which escapes our notice. The rocks, too, are mutinous with change, could our eyes but follow it. They too, grow, and change, and die, and give up their substances to the great All, returning whence they came. Changeless they seem, in contrast with the impatient waters; and yet with reluctant concession they give up their elements to the ambient air, and to the confluent restlessness of water, gradually rounding off their angles, and softening their rugged asperities. Mysterious and beautiful law, which ordains that the stubborn skeleton shall take its moulding from the gentle pressure of the softer flesh, as the sterner asperities of life are moulded finally by tenderness and love.

The Giant's Castle—indeed, the whole of this southern shore—has a character of drear magnificence and massive grandeur, given to it by the disposition of its piled-up boulders and towering altitudes, not to be anticipated from the size of the islands. The truth is, we are always impressed by relative, not absolute size. Rocks, many thousands of feet in height, have a stupendous aspect only in isolation; among others, of kindred girth and altitude, they produce no such towering impression. The eye takes its standard from the forms around. The subtle influence of proportion rouses emotions of the sublime, even on these small islands; emotions of gentler swell are raised by every creek and valley.

The rambles are delicious. They want, indeed, the charm of Devonshire, with its wondrous lanes—

> "Such nooks of valleys, lined with orchises,
> Fed full of noises by invisible streams." *

There are no rills and rivulets intersecting the land, no affluence of vegetation making it a miracle of beauty and of life; but the lanes have their charm, and to that charm I yielded myself.

After my first walk had satisfied the first cravings, and set the mind at ease respecting the wisdom of my choice in choosing Scilly, I returned to my lodgings, unpacked the book-box, arranged the working-table with its necessary jars, bottles, dissecting-implements, and microscope; and, resting from these labours, opened Borlase, to gain from his ancient quarto some

* *Aurora Leigh.*

information about the place. I will not, as some learned pundits do, pitilessly burden you with knowledge recently obtained; because, although I suspect you to be hopelessly ignorant on all these matters, I also suspect you to be quite comfortable in that condition, and by no means hungering for information; and at any rate, you know where such hunger can be satisfied. But on the baptism of the islands a word may be worth hearing. Borlase pertinently asks, "How came all these islands to have their general name from so small and inconsiderable a spot as the isle of Scilly, whose cliffs hardly anything but birds can mount, and whose barrenness would never suffer anything but sea-birds to inhabit there? A due observation of the shores will answer this question very satisfactorily, and convince us, that what is now a bare rock, about a furlong over, and separated from the lands of Guel and Brehar about half a mile, was formerly joined to them by low necks of land, and that Treskaw, St Martin's, Brehar, Samson, and the rocks and islets adjoining, made formerly but one island." Thus it was by encroachments of the sea, according to Borlase, or by the dipping of the lands, that the one island was separated into several. Scilly was the highest and most conspicuous headland, and from it the whole group derived its name. That these isles were by the Greeks called *Cassiterides*, and by the Romans *Sigdeles*, *Sillinæ*, and *Silures*, may be conceded to antiquarians and topographers, or denied; we shall trouble ourselves but slightly with the question. Certain it seems that Phœnicians and Romans came here for tin; still more certain that, in the tenth century, " when trade began to thrive, shipping to increase, and naval wars to be carried on in the western world, the commodious situation of these islands at the opening into both the Channels soon showed of what importance it was to possess them, and how dangerous they might be to the trade and safety of England if in an enemy's hand." The hungry may find in Borlase a succession of historical dates and facts from the tenth century downwards; we will pause only at what is said of Queen Elizabeth, who saw the importance of these islands; " and having the Spaniards, then the most powerful nation by sea in the world, to deal with, ordered Francis Godolphin (knighted by her in 1580, and made Lord-Lieutenant of the county of Cornwall) to improve this station. Star Castle was begun and finished in 1593. At the same time were built a curtain and some bastions on the same hill." The castle still remains; and the fortifications—not of a very formidable aspect—manned by five invalids, still keep up the fiction of awing the enemies of England. Not being a military man, and still less a politician, it does occur to me either that Scilly is strangely neglected in the matter of fortifications, or else that our enemies are very easily awed. What Borlase said of it a century ago remains true to-day : " In the time of war it is of the utmost importance to England to have Scilly in its possession : if it were in an enemy's hand, the Channel trade from Ireland, Liverpool, and Bristol to London and the south of England could not subsist;

for Scilly, lying at the point of England, and looking into both channels, no ship could pass, but a privateer might speak with it from one of these sounds. This the Parliament ministry in the latter end of the civil wars of Charles I. quickly experienced as soon as Sir John Granville had garrisoned and fortified Scilly. Whitelock tells us that continual complaints were made to the then managers of affairs at London, of the taking of ships by the privateers at Scilly, so that at last they were obliged to send Admiral Blake and Sir George Askue to dislodge the cavaliers from a post which gave them such opportunities of distressing their trade." Surely a post of this importance needs a stronger garrison than five invalids? Five may do for the "contingent" of a small German prince; nay, in one sublime instance, five is the sum total of the standing army, but in that case the principality itself is of commensurate importance.

What has been already hinted will suffice to show that these patches of rock, on which ribald Cockneys doubted whether English were spoken, and flounces worn, are islands dignified by historical and political associations. These Cockneys may be further assured that not only is English spoken here, but spoken with a purity of accent, and intelligent discrimination of diction, which I remember in no other part of the English dominion. The Scillians are a remarkably healthy, good-looking race—the black eyes and long eyelashes of the children making one's parental fibres tingle with mysterious pleasure as the ruddy rascals pause in their sport to look at the stranger. Their manners are gentle and dignified; civil, not servile. Not an approach to rudeness or coarseness have I seen anywhere.

In the highest sense of the word civilisation, therefore, the notion of the place being "half-civilised" is altogether wrong. It is only on making inquiries in the direction of commerce that the mind gets familiarised with the consequences of the remoteness of these islands. Then it is seen that, as far as civilisation is represented by shopkeeping, Scilly is at present in an embryonic condition. To speak physiologically, there is but slight *differentiation* of function in the Scillian commercial tissue. Just as in the simpler organisms we see one part of the body undertaking several functions which in more complex organisms devolve upon separate parts, so here we perceive the same smiling individual weighing out butter and measuring yards of muslin, proposing the new cut of a cheese to your discriminating taste, or the new style of bonnet to your instincts of fashion; sarsenet ribbons are flanked by mixed pickles, and the pickles thrown into relief by loaves. If you are troubled with a raging tooth, you must apply to the postmaster for his gentle services; whether he punches it out with the letter-stamp, or employs more elaborate instruments, I know not. This want of differentiation is, however, but a slight obstacle, especially to me, who am not likely to array myself in sarsenet, and don't buy bonnets. Far otherwise is the imperfection there where it could least have been expected, least endured—in the meat and market departments. It is probable, on zoological grounds, that the Scillians, being carnivorously organised, would eat meat with gusto could they get it. Nay, as there are several well-to-do people residing here, some shipowners and shipbuilders, and as there are no poor, it would, on *a priori* grounds, be assumed that meat was freely assimilated by the Scillians, they not having fallen into the fallacy of "vegetarianism." But *a priori* conclusions force no pathway through facts; and the stern and startling fact early obtruded itself on me, that of all things meat is one of the most unattainable in these parts. Do not imagine that by "meat" I euphuistically indicate prime parts, and quick varieties; no, I mean meat of any kind, without epicurean distinctions. Beef *is* obtainable—by forethought and stratagem; but mutton is a myth. A vision of veal floats with aerial indistinctness through the Scillian mind. Poultry, too, may be had—at Penzance; and fish—when the weather is calm, which it never is at this season; and when the one solitary fisherman adventurously takes out his line—which he seldom does. But market there is none. Twice a-week a vegetable cart from "the country" (which means a mile and half distance) slowly traverses the town, and if you like to gather round it, as the cats and dogs do round the London cats'-meat-man, you may stock yourself with vegetables for three days. The inhabitants, of course, know how to arrange matters for themselves, although it was evident that my landlady regarded the wish of dining daily, and if possible on meat, as rather a metropolitan weakness, which was to be politely allowed for. The other day I should have gone meatless, but for a certain astuteness of forethought, met by a yielding benevolence on the part of the captain's wife. Meat was not to be had for love or money, especially love. The " country " had been scoured for a fowl—

> "But no such animal the meadows cropp'd."

I saw myself midway in the dilemma of going *impransus*, or of cooking my Actiniæ with what appetite I could—an extremity which, in a zoologist, would have been only a milder form of cannibalism. Standing thus at the point of intersection of two such paths, the pangs of prospective hunger developed in me new resources and new impudences. I went boldly to Mrs Tregarthen (observe she is not a *widow*), and to her pathetically unfolded the case, on the supposition that she might not be utterly meatless, in which circumstance the loan of a chop or steak might gracefully be accorded. Meatless the gentle and generous woman was not. A piece of beef, killed eight days ago, and now kept fresh in salt against emergencies, would furnish me with a steak sufficient for two days, and there was a rumour that on the third day beef would be killed, when I could stock myself till next killing-time. Beef, at sevenpence a-pound, as I said, is the only meat you can reckon on, even with forethought In the time of Borlase it was just the contrary, mutton being then the meat, and beef a rarity. " About twenty years since," he says, " the inhabitants generally lived on salt victuals, which they had from England or Ireland; and if they killed a bullock here, it was so seldom, that in one of the best houses in the islands they have kept part of a bullock killed in September to roast for their Christmas dinner." He adds, that in his time mutton was abundant enough, but beef unattainable.

Spiritual-minded persons, indifferent to mutton, may disregard this carnal inconvenience, and take refuge in the more ideal elements of picturesqueness, solitude, and simplicity. I cannot say that the inconvenience weighed heavily in the scale against the charms of Scilly; the more so, as an enlarged experience proved the case not to be *quite* so bad as it seemed at first. After all, I came not here for sumptuous larders, but for zoological delights; and *those* were not wanting. Was not the mere aspect of the sea a banquet? Xenophon tells us that, when the Ten Thousand saw the sea again, they shouted. No wonder. After their weary eyes had wandered forlorn over weary parasangs of flat earth, and that earth an enemy's, wistfully yearning for the gleams of the old familiar blue, they came upon it at last, and the heart-shaking sight was saluted by a shout still more heart-shaking. At the first flash of it there must have been a general hush, an universal catching of the breath, and the next moment, like thunder leaping from hill to hill, the loosened burst of gladness ran along the ranks, reverberating from company to company, swelling into a mighty symphony of rejoicing. What a sight, and what a sound! There was more than safety in that blue expanse, there was more than loosened fear in their joy at once again seeing the dear familiar face. The sea was a passion to the Greeks; they took naturally to the water, like ducks, or Englishmen, who are, if we truly consider it, fonder of water than the ducks. We are sea-dogs from our birth. It is in our race—bred in the blood. Even the most inland and bucolic youth takes spontaneously to the water, as an element he is born to rule. The winds carry ocean murmurs far into the inland valleys, and awaken the old pirate instincts of the Norsemen. Boys hear them, and although they never saw a ship in their lives, these murmurs make their hearts unquiet; and to run away from home, "to go to sea," is the inevitable result. Place a Londoner in a turnip-field, and the chances are that he will not know it from a field of mangold-wurzel. Place him, unfamiliar with pigskin, on a "fresh" horse, and he will *not* make a majestic figure. But take this same youth, and fling him into a boat, how readily he learns to feather an oar! Nay, even when he is sea-sick—as unhappily even the Briton will sometimes be—he goes through it with a certain careless grace, a manly haughtiness, or at the lowest a certain "official reserve," not observable in the foreigner.* What can be a more abject picture than a Frenchman suffering from sea-sickness—unless it be a German under the same hideous circumstances? Before getting out of harbour he was radiant, arrogant,

---

* "Had a furious gale off Flamborough Head; saw many a dandy's dignity prostrated by sea-sickness; was sick myself, *but managed to keep it secret*."—HAYDON'S *Journal* (*Life*, iii. 62.)

self-centred; only half an hour has passed, and he is green, cadaverous, dank, prostrate, the manhood seemingly spunged out of him. N.B.—In this respect I am a Frenchman.

At the sight of the sea the Ten Thousand shouted. At that sight I too should have shouted, had not the glorious vision come upon me through the windows of a railway carriage; where my fellow-travellers, not comprehending such ecstasy, might have seized me as an escaped lunatic. But if my lungs were quiescent, my heart shouted tumultuously. *There* gleamed once more the laughing lines of light, *there* heaved and broke upon the sands the many-sounding waves; and at the sight arose the thought, obvious enough, yet carrying a sort of surprise, that even thus had the sea been glancing, dancing, laughing, breaking in uninterrupted music, ever since I had left it. While I was bustling through crowded streets, amid the " fever and the stir unprofitable," harassed by printers, bored by politicians,

" The weary, weary A, and the barren, barren B,"

bending over old books, engaged in serious work and daily frivolous talk, through all these hurrying hours, the tides had continued rising and receding, the pools had been filled and refilled, the zoophytes had quietly dedicated their beauty to the sun, the molluscs had crawled among the weeds, the currents of life had ebbed and flowed in the great systole and diastole of nature.

By a mysterious law, every Thirst blindly, yet unerringly, finds its way to the fountain. My thirst had led me here, to the shores of that ocean which Homer, "the paragon of philosophers," as Rabelais calls him, very unphilosophically styles "unfruitful," ἀτρύγετος. Barren it may have been to him, poor fellow, unable to use the microscope (he was blind, you know !) yet even he had intellectual vision enough to see that it was μεγακήτης, "abounding in marvels;" and he was not a man to pause open-mouthed at a slight deviation from ordinary appearances, as may be gathered from this single example: when Helen passes through the gates of Troy, under the eyes of Ucalegon and Antenor, those venerable and inspired men are by Homer seen to be "like cicadæ chirping on the trees"—surely a very strange phenomenon?—and as if this were not enough, their chirp is said to have a *lily-like* sound — ὄπα λειριόεσσαν — surely a strange intonation? If, therefore, to Homer, familiar with sights and sounds so unusual, the sea could nevertheless be held as abounding in marvels, judge what it abounds in for our more easily astonished minds.

Come with me to the rocks, on my first visit after arrival. The tide is not a very good one, but in a few minutes we discover that we are in the land of marvels. Here are the snaky-armed *Antheas* in abundance: green with ravishing pink tips; brown with silver-grey tentacles; and a few of quaker drab. Presently a noble *Crassicornis* reveals himself in a cleft—impossible to get at, unfortunately. But in few minutes another, then another, then a group, at last such quantities of them make their appearance, that the heart palpitates at the wealth promised by this abundance; and is not abundance of animals worth a few hours' discomfort on board the packet? Nay (now that it is past), what *was* that discomfort? A hurricane of blows upon the chisel answers with contemptuous emphasis.

It is laborious work this chiselling away of Anemones from the granite. The grey-slate of Ilfracombe was troublesome; the limestone of Tenby worse; but this granite opposes us with quite another stubbornness, and needs energetic patience to overcome it. In spite of March winds I am forced to take off my coat after a little of this hammering; and during summer heats the exercise would create a vapour bath, giving unpleasant extension to the faculty of perspiring, which is exerted by the twenty-eight miles of tubing (such is the calculation) possessed by our skins. After filling our baskets with as many of these Anemones as satisfy present desires, we begin turning over the stones. Presently we descry two specimens of marine spiders, or daddy-long-legs (*Nymphon gracile*), very curious to behold. They have no body to speak of; a mere line, not thicker than one of their legs, representing the torso. Tie a piece of silk thread, about one-fourth of an inch long, into four equidistant knots, and that will represent the body; from each of these knots let much longer pieces of the same thread dangle, and you have the legs; split the tip of the thread into three filaments, and you have the head; gum bits of dirty wool, about as large as a pin's head, on the second legs, and you have the egg-sacs: and with this the animal is complete.

The microscope reveals fresh wonders, the head being furnished with crab-like nippers; the alimentary tube, instead of occupying an isolated and dignified position in the body, meanders out into each of the legs, so that the leg repeats the body in its internal structure, as well as in aspect. This ramified alimentary canal is covered with brownish yellow globules or cells, called "hepatic cells," and supposed to represent a rudimentary liver. Mr Gosse, in his pleasant book on *Tenby*, mistakes this intestine for the circulating system; but the animal has *no* circulating system whatever. "Each of the long and many-jointed limbs is perforated by a central vessel," he says, "the walls of which contract periodically with a pulsation exactly resembling that of a heart, by which granules or pellucid corpuscles of some sort or other are forced forward." It was *food* which Mr Gosse saw thus moved; the blood-circulation, such as it is, he correctly saw in what he describes as the *extra-vascular* circulation; only we should add, that *vascular* circulation there is none. The blood, if blood it can be called, is *outside* the intestine, bathing the walls of the body, and moved to and fro by the peristaltic action of the intestine. Curious as this *Nymphon gracile* is, I had reason to be the more pleased at finding one, because while the latest authorities declare nothing to be known of the development of the *Pycnogonidæ*, I had been fortunate enough, at Ilfracombe, to discover some of the embryonic phases, of which I made drawings, and awaited further opportunity for pursuing the subject.

If the reader will turn to Plate V., fig. 3, he will see a *Nymphon gracile* of natural size ; and having marvelled at its aspect, will marvel still more at fig. 4, which is a magnified representation of the same animal in the egg; and he may puzzle himself by trying to conceive the stages of metamorphosis through which this creature must pass before it can exchange the squab rotundity of its early shape for the slim and meagre elegance of maturity. Are Sylphs bulbous in early life? By what modifications could a crab be supposed to pass into a daddy-long-legs? Yet some such changes must take place with the *Nymphon gracile*, although I have not been fortunate enough to trace them beyond the point represented in fig. 4. But this is a digression; we must continue our hunting.

Here, in a pool, we find three curious fish, one a ribbon-fish, the other two unknown to me; and on raising the stone, behold, a queer eel-like fish, with a miniature grey-hound's head; it is the pipe-fish, *Syngnathus anguineus* (Plate VI., fig. 1). Pop him in ; also this bit of red weed, on which I observe some *Polyzoa* clustering.—(Plate I., fig. 1, represents a fresh-water Polyzoon, named *Plumatella*, expanded and contracted.) What is this? a tiny Daisy on a frond of weed? the beauty! No, now it is in the bottle, it turns out to be an Eolis, *Eolis alba*, lovely among the loveliest. Stay! here are two *cowries*, and alive! The shells every one has seen, but few of us have seen the animals ; so the capture is very welcome. My back is aching with all this stooping and groping and I really must get home now, content with my day's work. One farewell glance in at that pool, and I have done. Lying on my face, and dangling my feet in water, I peer scrutinisingly for some minutes, and bear off a lovely green *Actæon* as a reward. (The Actæon resembles an Eolis without the papillæ.) Now I *will* turn homewards.

Another day, in idler mood, we ramble along the shore in receipt of windfalls. A bottle is always ready in the pocket, and something is certain to turn up. The stem and root of that oar-weed, for example, is worth an investigating glance, certain as it is of being a colony of life. The tiny Annelids, white, green, and red, wriggle in and out among the sheltering shadows of these roots; the Sponges and Polyzoa cluster on them; and see! what pink-and-white feathery creature is this, clasping the weed with a circle of pale pink roots? By heavens! it is a *Comatula* (Plate VI., fig. 2); and now that it feels the grateful sea-water again, how it expands its feathers, and reveals itself as an animal fern, marvellous to look upon. Sudden joy leaps in our hearts at the sight of this creature, hitherto known only from hazy descriptions and inadequate engravings. There is interest in reading about *Crinoidea*, fossil and recent, and in learning that the *Comatula* is one of these, having kindred with starfishes; but how that interest is intensified by direct inspection of the living animal! I could not satiate myself with looking at my prize.* All the

---

* I have since had several, but utterly inferior in colour and grace to this, the first I ever saw.

way home the bottle was constantly being raised to my loving regard, that I might feast myself upon the waving grace of those pink-and-white feathers; and I thought of the poetical passage in which Edward Forbes expresses his emotions about these *Crinoidea* which "raise up a vision of an early world, a world the potentates of which were not men, but animals— of seas on whose tranquil surfaces myriads of convoluted Nautili sported, and in whose depths millions of Lily-stars waved wilfully on their slender stems. Now, the Lily-stars and Nautili are almost gone; a few lovely stragglers of those once abounding tribes remain to evidence the wondrous forms and structures of their comrades. Other beings, not less wonderful, and scarcely less graceful, have replaced them; while the seas in which they flourished have become lands whereon man in his columned cathedrals and mazy palaces emulates the beauty and symmetry of their fluted stems and chambered cells."*

The delight of getting new animals is like the delight of childhood in any novelty, an impulse that moves the soul through the intricate paths of knowledge—knowledge which is but broken wonder; and this delight the naturalist has constantly awaiting him. Satiety is not possible, for Nature is inexhaustible. Knowledge unfolds vista after vista, for ever stretching illimitably distant, the horizon moving as we move. New facts connect themselves with new forms; the most casual observation often becomes a spark of inextinguishable thought, running along trains of inflammable suggestion.

* *History of British Star-fishes*, p. 2.

To this intent the naturalist should always have pencil and note-book on his working-table, in which to record every new fact, no matter how trifling it may seem at the moment; the time will come when that and other facts will be the keys to unlock many a casket. Not that Observation alone is, as many imagine, the potent instrument of Zoology. Lists of details crowd books and journals, yet these are in themselves no better than the observations of Chaldean shepherds, which produced no Astronomy in centuries of watching. They find their place in science, only as the architectural mind disposes them in due co-ordination. What should we think of a chemist who, on mere inspection of substances, unaided by re-agents and his balance, hoped to further Chemistry? What would lists of such observation avail? And in the far more complex science of Biology, how shall cursory inspection, superficial observation, avail? We must follow the Methods which have led to certainty in the exact sciences. We must render the complex facts of Life as simple as we can, by processes of elimination. Experiment must go hand in hand with Observation, controlling it, and assuring us that we have correctly observed. Much has been done, and is daily done, in this way, yet still men too easily content themselves with observation, or, what is equally fallacious, with anatomical deduction, declaring an organ to have such and such a function, merely because it resembles an organ known to have the function; when in most of these cases, direct experiment would show the error of the conclusion.

In former chapters I have illustrated this point, and have again to do so respecting the digestive power of the Sea-Anemones.

In my note-book is pencilled this brief query, "Do the Actiniæ digest at all?" a doubt which, in its naked simplicity, might rouse contempt in the mind of any zoologist accidentally reading it. What! here is an animal notoriously carnivorous, and you ask whether it can digest? Have not you yourself repeatedly *fed* these animals with limpets and cooked beef? are they not greedy of such food? It is perfectly true. Nevertheless a doubt occurred to me whether they did really digest, in any proper sense of the term; and I made a note of the doubt, as of a point to be investigated immediately on my arrival at the coast. Experiment should settle the doubt. Before narrating the experiments, it will be needful to settle with the reader a few generalities on the subject of Digestion; since, in point of fact, the interest of the question falls mainly on the general subject, and only with a secondary importance on the digestive powers of the Anemones.

What are we to understand by Digestion? At first the question seems so easy; yet the closer it is investigated, the remoter seems the possibility of answering it. Let us make a clearance by first discriminating Digestion—as a *special function* of the intestinal canal—from Assimilation, which is the *general property* possessed by all living tissues. For an animal to grow, and to repair the waste which the action of life incessantly produces, it must assimilate, which, as the word implies, means to separate from the external medium such substances as are *like* to its own substance—or can be converted into them by the vital chemistry—rejecting all such as are unlike, or not convertible. Very simple organisms find assimilable food in the element they live in, and the process of separation is easy: they have no stomach, not even a mouth, much less glands secreting solvent fluids. Very complex organisms, on the contrary, do not, in the air they breathe, or on the earth they tread, find the variety of substances necessary to build up their bodies; the substances have to be sought, captured, and when found, are not found in an assimilable condition, but in a condition requiring great changes, mechanical and chemical, before they are able to enter into the construction of the tissues.

An example will make this plain: Let us first consider the process in the *Actinophrys*, a microscopic animal carefully studied by Kölliker.* It is a mere mass of jelly-like substance, very contractile, without the slightest trace of organs, perhaps also without even a distinct envelope separable from the mass.†

---

* Siebold ü. Kölliker's *Zeitschrift für Wissenschaftliche Zoologie*, i. 198.

† On the existence of a distinct membrane there is much dispute. The arguments, for and against the existence of single-celled animals, are such as, in the present state of inquiry, render decision difficult. To my mind, however, the researches of Auerbach (*Ueber die Einzelligkeit der Amoeben*, in Siebold ü. Kölliker's *Zeitschrift*, vii. 365) render extremely probable the fact that an enveloping membrane and a nucleus always exist, which afford a very strong argument in favour of the unicellular structure of these animals. Although I have been unable, in repeated examinations, to convince myself of the existence of a

The outer layer is formed into long tentacular filaments, which, like the tentacles of a Polype, seize hold of young animalcules, or even minute Crustaceans. The resemblance to the Polype is carried further: no sooner does one of the filaments seize a prey than it retracts; all the others bend their points over the captive, and gradually enclose it; they then retract, and bring the food in contact with the body of the animal. The point of contact is next seen to yield inwards, retracting as the filaments had retracted, and, as it deepens, the food sinks into the substance of the body, the edges of the cavity closing over it. In the centre of the body the soluble parts are dissolved, the body having resumed its original appearance. This done, the insoluble parts make their way out, much as they made their way in; and thus the whole process of ingestion and egestion is accomplished.

We need not pause to trace the episodes of the complex story of Digestion in the higher animals, episodes of mastication, insalivation, chymification, chemical transformations aiding mechanical actions; every one is familiar with the general facts. Let us only note

distinct membrane, optical delusion being easy in this matter, I cannot resist the conclusion that something of the kind does exist, inasmuch as there are currents inside the body of the *Amœba*; and after watching these currents, or rather *oscillations*, carrying the large dark globules with them into the various prolongations of the body and back again, for upwards of an hour, I felt convinced that the Amœba was a cell, or at any rate a closed cavity with liquid contents. As regards what is said in the text, the point is of no importance; and this note is mainly added for the sake of directing the reader's attention to AUERBACH's interesting essay.

that even milk, which contains all the substances needed for the nourishment of the child, contains them in a condition perfectly useless, as far as the direct and immediate nourishment of the child is concerned: until the milk has undergone the digestive process—namely, a succession of chemical decompositions and recompositions—it is no more competent to nourish the muscles, bones, and nerves of the child, than so much chalk-and-water, which is delusively sold as milk in virtuous cities. The mutton-chop, too, which we justly reckon such excellent food, is only "food potential;" it must undergo a very curious series of changes before it can be converted into blood. Nor is the business finished there. We are erroneously accustomed to consider blood as the final stage of food, previous to its assimilation. Physiologists trace the story of Digestion up to this point, and there leave it; as story-writers leave their heroes married, thereby indicating that nothing more remains to be said. But just as marriage is the beginning of a new act in the drama, and the act in which all life culminates, so is this blood-formation but the commencement of a new series of changes, and these the most important. I think it can be shown that the blood itself is not more immediately and directly assimilable than the mutton-chop from which it was formed. In its passage through the walls of its vessels, it undergoes specific changes, fitting it for assimilation; without such changes it is not assimilable; blood, *as* blood, nourishes no tissue, but lies on it like any other foreign substance which must be got rid of by re-absorption into the veins—as we see when a vessel is ruptured, and the blood gets deposited in the parenchyma. Blood is, in fact, as Bergmann and Leuckart well express it, "a depôt of assimilable and secretory substances; and its purpose in the economy is that of a regulating apparatus, which is necessitated by the fluctuations in the procuring of food."*

Remember also, that, before Assimilation can take place, the food must be rendered soluble. Solubility is a primary condition, but not the only one. Many soluble substances have to undergo chemical changes, both of decomposition and allotropism, before they form parts of the living body. If albumen or sugar be injected into the veins, they will not be assimilated, but cast out unaltered in the excretions; whereas, if injected into the alimentary canal, or into the portal vein, which would carry them through the laboratory of the liver, they are entirely assimilated.

Thus we see that solubility and transformation are the two digestive effects, to produce which, two agencies are needful, the mechanical and chemical. From these two points all other questions expansively radiate, to them all converge. A single fact strikingly impresses the mind with a sense of the extent to which chemical agency reaches, namely, that in the course of four-and-twenty hours a sixth part of the whole weight of the body is poured into the alimentary canal, under

* BERGMANN ü. LEUCKART: *Vergleichende Anatomie und Physiologie*, p. 164.

the form of various secretions. Much more fluid is secreted from the blood and poured into this canal during a single day, than would make up the whole mass of fluid circulating in the blood-vessels at any given period.*

The reader's attention has been so fully directed to this twofold agency of Digestion, and especially to its chemical agency, that a clear view may be taken of the question which must arise as to what, in the abstract, is the purpose of Digestion. In the abstract we may declare it to be the *preparation* of the food, rendering it fitted for Assimilation. But if we descend from heights of abstraction, and approach concrete questions, we soon find this answer including several processes in the higher animals—such as the prehension and mastication of food, its absorption and circulation, its aeration in the blood, and finally its transudation through the walls of the capillaries—none of which can, without great impropriety, be called *digestive*. We must be more specific. No man would confound Mastication with Digestion, or Circulation with Digestion; and we must therefore limit the term Digestion to some specific meaning; Mastication is the special function of the jaws, Circulation of the vessels, Respiration of the lungs, and Digestion of the alimentary canal. But even this is too vague for our purpose; we must affix a still more specific character to Digestion; and this may be expressed in the following formula: That, and that only, is a specifically

* LEHMANN: *Lehrbuch der Physiol. Chemie*, iii. 226, 2d edit.

digestive act which takes place in an alimentary canal, by means of secretions capable of *chemically* modifying the food, so as to prepare it for Assimilation.

The preparation of food we have seen to be both mechanical and chemical, but I select the latter as the *specific* characteristic of the digestive process, in order to prevent confusion. Claude Bernard says: "We can conceive an animal without any digestive apparatus, mechanical or chemical, because living in an element which furnishes nutritive material *directly;* we can also conceive the digestive act reduced to a simple mechanical apparatus, which has to press out certain alimentary juices capable of nourishing the tissues *without undergoing chemical modifications;* but usually the digestive act is composed of two orders of phenomena, physical and chemical."* This is a brief and luminous classification as regards the whole animal series, and it well expresses the ascending complexity of that series; but inasmuch as special functions only make their appearance at certain stages of that ascending series, inasmuch as the simpler animals have not the special functions of more complex animals, we must deny to the two first classes of M. Bernard's series any such special function as Digestion, and confine it to the third class. We do not, except in loose latitude of phrase, speak of the *legs* of an animalcule, meaning its organs of Progression; because a leg is a specific organ of Progression, uniform in its elements throughout the series of animals possessing legs; nor

* CLAUDE BERNARD: *Leçons de Physiol. Expérimentale*, ii. 490.

should we, otherwise than in easy speech, talk of the Digestion of a Polype, meaning thereby its Nutrition. The *purpose* of a leg, progression, is fulfilled by the cilia, which move the animalcule; the *purpose* of Digestion, preparation of food, is performed by the cavity of the Polype; but the specific organs named legs and alimentary canal, and the specific functions of those organs (Walking and Digestion), are in both cases absent.

If the reader has followed me thus far, he will have understood that, when I doubted whether the Actiniæ digested, there was no doubt entertained of their power of preparing food, but only of their power of *chemically* digesting it. I doubted, in short, whether they should not be separated from the more complex animals which digest, and whether they should not rank in M. Bernard's second class. We do not call a hut or group of cottages a city. We do not speak of its commerce, its government, its literature; these are social functions developed in a complex city, not possible in a group of cottages. In the same way we should not expect to find Digestion, Respiration, or any other complex function, in animals so simple as a Sea-Anemone. Nor could the notion ever have gained currency, had there been the proper precision in our zoological language, and had not the "fallacy of observation" misled us.

Now to the experiments. The first point to be settled was this: Have the Polypes anything of the nature of a solvent fluid secreted by their stomachs? "It is obvious," says Dr Carpenter, the latest writer on this subject, "that a powerfully solvent fluid is secreted from the walls of the gastric cavity; for the soft parts of the food which is drawn into it are gradually dissolved, and this without the assistance of any mechanical trituration." Obvious, indeed, the fact seems, until it is interrogated a little more closely, and then we find, 1st, that *no* solvent fluid is secreted; 2d, that the food is not dissolved, but only the juices pressed out.

My first experiment was to test the presence or absence of a secretion, which was accomplished thus: Tying a narrow strip of litmus-paper round a small piece of recently caught fish, and fastening it to a thread, I gave it to an *Anthea cereus*, who greedily swallowed it; another thin slice of the same fish was folded longitudinally over a similar bit of litmus-paper, and given to a *Crassicornis*. If any acid secretion were present, the paper would redden; if not, the blue colour would remain. On the following morning the ejected morsels were examined, but not a trace of acid reaction was visible. Repeating the experiment several times under varying conditions, I came to the conclusion that no acid fluid is present in the digestive process of the Actiniæ. There still remained a doubt. Solvent secretions are either acid or alkaline. It was necessary to make similar experiments with an alkaline re-agent. This was done, and with similar results.

It is worth noting that M. Hollard equally failed in detecting an acid or alkaline reaction,* which is a confirmation of my experiments.

The Actiniæ do not effect their preparation of nutriment by *chemical* means; and in our strict sense of the term, they cannot be said to digest. I was anxious to see how far *mechanical* means were employed, and for this Réaumur's admirable experiment was a guide. In his day it was supposed that digestion was a purely mechanical operation, the food being ground into a pulp in the stomach. He took hollow silver balls, perforated with holes, and filling them with meat, caused them to be swallowed by a dog. When they had remained a suitable period in the animal's stomach, they were withdrawn by the thread attached to them. If the digestive process were mechanical, the meat would be protected from all grinding action, by the silver covering; if chemical, the meat would be digested; and digested (or rather chymified) it proved to be; showing that a solvent fluid had penetrated the holes, and dissolved the meat. I took two pieces of quill, of about half an inch in length, open at both ends, and having six good openings cut in the sides, thus affording ample means for any solvent fluid to exert its action on the roast-beef enclosed in the quill. On examination of the ejected quills, I found no ap-

* "Il est remarquable, et je m'en suis souvent assuré, que les papiers réactifs plongés dans cet organe, et dans la cavité inférieure, soit au moment de la digestion, soit chez l'animal à jeûn, ne donnent aucun indice d'acidité, ni d'alcalinité."—"Etudes Zoologiques sur le genre Actinia."—*Revue et Magazin de Zoologie*, No. 4. 1854.

preciable difference between the contained meat, and similar pieces of meat left in the water during the same period; in one of them which had the meat protruding somewhat from each end of the quill, there was a maceration of the protruded ends, which looked like a digestive effect; but on submitting it to the microscope, I found the muscle-fibres not at all disintegrated, the striæ being as perfect as in any other part, and the maceration obviously of a purely mechanical nature. A similar appearance is presented by meat, after its ejection by the Actiniæ: it is pulpy, colourless, but the muscles are not disintegrated.

Mr R. Q. Couch, of Penzance, was good enough to repeat these experiments for me. "I folded portions of whiting in test-papers," he writes, "and gave them to the Actiniæ. After 12 hours the whole was ejected without the papers being either broken or discoloured. I placed bits of mackerel in gutta-percha silk with the same result. Taking other specimens, which I kept fasting for a fortnight, I gave each a portion of the silvery part of a mackerel, measured its length, and weighed it in a delicate balance. In one case the fish was ejected after 23 hours, and in another 32 hours, and in several others about 18 and 20 hours. The parts were folded on each other, and compressed into oval masses which, with the point of a pen, I could unravel easily. In measure they were precisely as I had given them. In weight there was a difference: in one case, 9 grs. were reduced to $5\frac{1}{4}$ grs.; in others, 8 grs. to 5; 11 grs. to 5; 7 grains to $3\frac{1}{2}$.

In one piece from which I had expressed all the fluid I could, the decrease was slight, and the food was soon rejected. *But in every case the delicate skin of the ventral portions of the mackerel and whiting were uninjured.* The fine metallic lustre was untouched, and the skin unbroken; showing that the digestion does not consist in appropriating the substance of the food given, but in expressing the juices contained therein."

I dare not pause now to touch upon the many topics which are suggested by the conclusion to which these investigations led me. It will be enough just to note here the progressive complication of the digestive function in the progressive complexity of the animal series. Starting from the simple cell, which draws its nutriment from the plasma surrounding it, by a simple process of endosmosis, we first arrive at the mouthless *Actinophrys*, or the *Amœba*, which, folding its own substance over the food, presses out such nutriment as it can; we then reach the *Infusory* with a mouth, but without stomach of any kind;[*] and the *Polype*, which has a portion of its integument folded in, serving both for mouth and stomach, but not anatomically differing from the external integument, nor physiologically differing in its action from that of the *Amœba's* gelatinous substance;[†] we then ascend to

---

[*] Nobody now believes in Ehrenberg's *Polygastrica*, or many-stomached animalcules.

[†] Trembley turned a Hydra inside out, and found the outside perform the function of a stomach. This has been held as proof that a mucous membrane is only a reflection of the skin. But from what has

Annelids having a real intestine, lying free in the general cavity, but only moderately, when at all, furnished with secretory apparatus; and so on, till at length we reach the *Mammalia*, with their marvellously complex digestive apparatus. Corresponding with this increasing complexity of the organs is the increasing complexity of the food which the animals digest, from simple gases up to meat.

If all were not so marvellous in Nature, would not the marvellous fact that food itself exists, arrest us? Food is what the organism can separate from the world around it, converting what it separates into its own life. We may consider Life itself as an ever-increasing identification with Nature. The simple cell, from which the plant or animal arises, must draw light and heat from the sun, nutriment from the surrounding world, or else it will remain quiescent, not alive, although latent with life; the grains in Egyptian tombs, after lying thousands of years quiescent in those sepulchres, are placed in the earth, and then smile forth as golden wheat. What we call growth, is it not a perpetual absorption of Nature, the identification of the individual with the universal? And may we not in speculative moods consider Death as the grand impatience of the soul to free itself from the

---

been advanced in this chapter, the reader may suspect that, inasmuch as the polype has *no* mucous membrane whatever, the so-called stomach not being anatomically distinguishable from the external skin, and the process of digestion being wholly mechanical, the current opinion is not proved by Trembley's experiment.

circle of individual activity,—the yearning of the creature to be united with the Creator?

As with Life, so also with Knowledge, which is intellectual life. In the early days of man's history, Nature and her marvellous ongoings were regarded with but a casual and careless eye, or else with the merest wonder. It was late before profound and reverent study of her laws could wean men from impatient speculations; and now, what is our intellectual activity based on, except on the more thorough mental absorption of Nature? When that absorption is completed, the mystic drama will be sunny clear, and all Nature's processes will be visible to man, as a divine Effluence and Life.

CHAPTER II.

DRUIDICAL REMAINS—A WRECK OFF SCILLY—GEOLOGY AND ZOOLOGY OF THE ISLES—EFFECT OF LIGHT ON PLANTS AND ANIMALS: HISTORY OF ITS DISCOVERY—THE PIPE-FISH AND ITS INCUBATION—FISH PARADOXES—AN AQUARIUM—SUICIDE OF THE STAR-FISH—THE PLEUROBRANCHUS—DEVELOPMENT OF EOLIS, DORIS, AND ACTÆON—SHELL AND NO SHELL—THE PEDICELLINA: IS IT VIVIPAROUS?—THE SAGITTA: A PUZZLE TO ZOOLOGISTS—WHERE THERE IS NO RESPIRATION THERE WILL BE NO CIRCULATION—THE CHYLAQUEOUS FLUID OF ANEMONES PROVED NOT TO EXIST—EARLIEST STAGE OF A NUTRITIVE FLUID—FUNCTION OF THE "CONVOLUTED BANDS"—DELIGHTS OF LITERATURE.

THE traveller's first wish is Shakespeare's—

> "I pray you, let us satisfy our eyes
> With the memorials and the things of fame
> That do renown this city."

At Scilly there is no city, and this non-existent city boasts "no things of fame," unless we choose so to consider the grave where Sir Cloudesley Shovel was first interred—which crowns the negative attractions of the Isles by being no grave at all. I am quite serious. They ask you here, whether you have seen the grave; on investigation, this renowned spot turns out to be destitute even of the rudest stone or landmark to indicate where the bones of the wrecked admiral may imaginatively be supposed to lie; it is simply a strip of land on the coast, where no grass will grow by reason of the shifting sand. And yet, if "gossip report" be not wholly a fibber, somewhere in this neighbourhood lie the remains of the great admiral, who was wrecked as he returned home covered with glory, 1500 or even 2000 men perishing with him on these inhospitable rocks. This was a century and a half ago, and tradition, we know, is apt to magnify, *vires acquirit eundo*. Still, if they will keep up the tradition, they might put up a commemorative stone. Stones are abundant enough, in all conscience; and, if we believe the antiquaries, some of these stones are invested with the hoar of Druidical sanctity.

Druidical erudition is not common. On probing the recesses of my own knowledge on this mysterious subject, I found that the principal source of my familiarity with it was the opera of *Norma*. For more than twenty years I had reverently followed that splendid priestess Giulia Grisi, and that majestic priest Lablache; and if to these you add the fragments of undeniable Druidical Remains in the persons of the very ancient virgins of the sun, forming the nightly chorus of that opera, little doubt should be thrown on the accuracy of my historical conceptions. With that erudition I had been content. But on reaching Scilly, where the respectable Borlase assured me Druid temples and sacred rock-basins did veritably exist, I was not a little anxious to bring my operatic erudition into direct confrontation with fact. I even cleared my throat for a pathetic burst of *moriam insieme*, when I should really stand beside a Tolmen, and with the mind's eye behold my *casta diva* about to perish, the victim of a superstition which had small sympathy with lovers.

Following Borlase's directions, I soon came upon a towering altitude of stones, in solitary isolation on the shore. A less erudite eye would have seen here nothing but a pile of stones; but the forewarned mind descried in their symmetrical arrangement, ledge upon ledge, crag upon crag, the rude architecture of early days; especially when it glanced at the stone-hedges and stone-cottages near at hand, which assuredly *were* built by human architects, and showed a less symmetrical arrangement than the towering pile. Then, again, the rock-basins, in which the pure water of heaven was received, who could doubt that their oval form and smoothly chiselled sides and bottoms were the work of man? If the cairn of stones left vague doubts, these rock-basins veritably *were* Druidical remains; and thus fortified against scepticism, I indulged in the emotions which naturally accompanied the belief of being in the presence of the remnants of a great human epoch long since passed away.

Having indulged in these emotions, and extracted from them all the pleasure they could yield, it was with acquiescent equanimity that I afterwards learned how little probability historical scepticism allowed to these Druidical remains. It appears that the cairns are simply cairns, and not temples. The architecture is Nature's; and, indeed, the forms are repeated in almost every cairn along the shores. Moreover, those rock-basins, which looked so convincingly human in their design and execution, are proved by Science to be the result of the disintegrating action of winds and waters, the uniformity of the causes producing that uniformity of result which seemed the betrayal of design. There is something almost pathetic in an acute and erudite man like Borlase (a naturalist too, and inventor of the strange worm which bears his name, *Nemertes Borlasia*), wandering among these rugged rocks, and finding in them the traces of an ancient religion; noticing the oval basins, and believing them to be human work; inventing a plausible explanation of their uses, admiring their design, and feeling a sacred awe in their presence; whereupon arrives the geologist with his disintegrating explanation, and the whole erudite fabric falls to pieces. Had Borlase lived in our time, imagine the ineffable scorn with which he would have looked down upon my Druidical authority, *Norma*; yet, you see, he is, with all his learning, quite as unveridical as Giulia Grisi, and not half so beautiful. If *Norma* is not a good historical authority, it is at least a delightful one; and, with Voltaire, I exclaim—

> "On court, hélas, après la vérité;
> Ah! croyez-moi, l'erreur a son mérite."

Scepticism refuses admission to these Druidical remains altogether, so that I need not occupy space with the description of them. But here is a story safe from the assaults of scepticism, and thrilling enough it is to deserve a place among moving accidents.

On the 16th November 1840, the French brig *Nérine*,

under Captain Pierre Everdert, with a cargo of oil and canvass, sailing from Dunkirk for Marseilles, was forced to heave to in a gale about ten leagues south-west of the Scilly Islands. The crew consisted of seven, including the captain and his nephew, a boy of fourteen. At seven in the evening a heavy sea struck the vessel, and completely capsized her—*turning the keel upwards.* The only man on deck at the time was drowned. In the forecastle were three men, Vincent, Vantaure, and Jean-Marie: the two former, by seizing hold of the windlass-bits, succeeded in getting up close to the keelson, and so kept their heads above water. The unfortunate Jean-Marie probably got his feet entangled—at any rate, after convulsively grasping the heel of Vantaure for a few seconds, he let go his hold, and was drowned. "The other two, finding that the shock of the upset had started the bulkhead between the forecastle and the hold, and that the cargo itself had fallen down on the deck, contrived to draw themselves on their faces close alongside the keelson towards the stern of the ship, from whence they thought they heard some voices. At the time of the accident, the captain, the mate Gallo, and the boy Nicholas, were in the cabin. The captain caught the boy in his arms, under the full impression that their last moment had arrived. The mate succeeded in wrenching open the trap-hatch in the cabin deck, and in clearing out some casks which were jammed in the lazarette (a sort of small triangular space between the cabin floor and the keelson, where stores are generally stowed away): having effected this, he scrambled up into the vacant space, and took the boy from the hands of the captain, whom he then assisted to follow them. In about an hour they were joined by Vincent and Vantaure from the forecastle. There were then five individuals closely cooped together; as they sat, they were obliged to bend their bodies for want of height above them, whilst the water reached as high as their waists; from which irksome position, one at a time obtained some relief, by stretching at full length on the barrels in the hold, squeezing himself close up to the keelson." What a situation! To rightly conceive its horrors, we must know that their only means of distinguishing day from night, was by the light which struck from above into the sea, and was reflected up through the cabin skylight, and thence through the trap-hatch into the lazarette. "The day and night of Tuesday the 17th, and of Wednesday the 18th, passed without relief, without food, almost without hope; but each encouraged the others when neither could hope for himself; endeavouring to assuage the pangs of hunger by chewing the bark stripped off from the hoops of the casks. Want of fresh air threatening them with death from suffocation, the mate worked almost incessantly for two days and one night, in endeavouring, with his knife, to cut a hole through the hull."

There is something very terrible in contemplating such a position, in seeing the mad energy of the mate thus to cut a hole, which would have caused instant destruction to the sufferers, since it was solely owing to this confined air that the vessel floated. Bad as the tainted air was, and threatening life every hour, it was the sole safety of the crew. They knew nothing of this; and when the mate's knife broke, a savage wrath at their frustrated hope must have seized them. "In the dead of the night of Wednesday, the vessel suddenly struck heavily: on the third blow the stern dropped so much that all hands were forced to make the best of their way, one by one, further towards the bows; in attempting which poor Vincent was caught by the water and drowned, falling down through the cabin floor and skylight. After the lapse of an hour or two, finding the water to ebb, Gallo got down into the cabin, and whilst seeking for the hatchet which was usually kept there, was forced to rush again for shelter to the lazarette, to avoid being drowned by the sea, which rose on him with fearful rapidity. Another hour or two of long-suffering succeeded, when they were rejoiced to see, by the dawning of the day of Thursday the 19th, that the vessel was fast on the rocks, one of which projected up through the skylight. The captain then went down into the cabin, and found that the quarter of the ship was stoved; and, looking through the opening, he called out to his companions above, ' Grâce à Dieu, mes enfans ! nous sommes sauvés ! Je vois un homme à terre !' Immediately after this the man approached, and put in his hand, which the captain seized, almost as much to the terror of the poor man as to the intense delight of the captain. Several people of the neighbourhood were soon assembled; the side of the ship was cut open, and the four poor fellows were liberated from a floating sepulchre, after an entombment of three days and three nights in the mighty deep." There is another curious detail in this story which must not be omitted. On Wednesday afternoon, two pilot-boats fell in with the wreck floating bottom up, at about a league and a half from the islands. They took her in tow for about an hour, when their towing-ropes broke, and as night was approaching, with a heavy sea running and bad weather threatening, they abandoned her, not having the faintest suspicion that there were human beings alive on board a vessel which was floating with little more than her keel above water. Nevertheless, although they abandoned the wreck, their temporary aid had been essential; had they not taken her in tow, the set of the current would have drifted her clear of the islands into the broad Atlantic waste.*

Granite is the substance of these islands. Generally it is thought that Scilly is only a continuation of the granite of Land's End; against which conclusion the idea of a separate and distinct range seems supported by the fact that, in dredging between the islands and the mainland, sea-weed is often brought up attached to bits of slate and greenstone; and the Wolf Rock, which lies not far southward of a line from the Land's End to Scilly, is composed of this same greenstone. What geologists call "the strike" of the granite here is, with few exceptions, towards the north or north-north-west.

---

* For this narrative I am indebted to North's *Week in the Isles of Scilly*—a work full of valuable details for any one who may contemplate a visit to these Isles.

The rock itself is not always confined to the constituent parts of quartz, felspar, and mica: schorl is a very common ingredient, sometimes accompanying the mica, sometimes replacing it. Hornblende is rare, chlorite still rarer. Veins of pure white quartz, of considerable size, often intersect the granite; rose-coloured quartz, and even chalcedony, have been found; but the general nature of the stone is of a coarse kind, useless for quarrying; and the granite needed for the new lighthouse is brought from Cornwall.

The reader will be curious to know about the zoological wealth of Scilly. Rich the place undoubtedly is, yet not so rich as I anticipated. When Dr Acland, of whom Oxford is justly proud, commenced the foundation of that anatomical museum over which he presides, Scilly was the first place chosen by him for the collection of specimens, on account of its geographical position at the entrance of the Bristol, Irish, and English Channels, with Rennel's Current near. He employed Victor Carus, since known by an excellent work on Morphology,[*] as his purveyor for six months, exploring and dredging. Carus has contributed a little paper to North's *Week at Scilly*, in which he gives expression to his opinion that the " sea is not a dense one, although there are multitudes of zoophytes and hosts of fishes; there are only a few molluscs, some worms, and a not very large number of echinoderms." On the whole, he does not think Scilly equal to the Channel Islands. Either I have been lucky, or my

[*] *System der thierischen Morphologie.* 1853.

wishes pointed in different directions from those of Victor Carus; for although unable to dredge, and confined, therefore, to tide-pools, I have had an embarrassment of riches rather than a want thereof. His verdict, however, is worth remembering, because, as these Isles are very inaccessible, and are hyperborean in the imperfection of their commissariat, the naturalist should weigh advantages with disadvantages before coming here. The attractions are manifold, as I have before explained; but the attraction of a *very* rich fauna Scilly cannot boast, unless zoophytes be the main object of search. The Anemones are various, and prodigally abundant. *Anthea*, and the noble *Crassicornis*, are almost as frequent as the *Smooth Anemone* is at Ilfracombe and Tenby. *Gemmaceas* abound; *Daisies* are frequent; the *Dianthus* is to be had; also the *Orange-disked*; and two species, probably yet undescribed—of which more anon.

To learn the geographical position of Scilly—above all, to get a glance at the coast—you would imagine it to be a wonderful place for marine zoology. The first obstacle lies in the nature of the rock. Granite, indeed, as mere granite, is almost as bad as chalk cliffs, which let no ingenuous reader waste his holiday upon. The weeds are loth to grow there; and where no weeds grow, no herbivorous animals will congregate for pasture; consequently no carnivorous animals will be there to pasture on them. The large amount of silica in granite resists the decomposing action of winds and waves, and of course still more energetically resists the animals, who require, among other things, lime for their shells. Drear and barren is many a hopeful-looking reef here : and barren they would all be, were it not for the compensating conditions of climate and tidal current. Scilly is a little to the west of the sixth degree of western longitude, and exactly in the fiftieth degree of northern latitude ; consequently it is the most southern part of the United Kingdom, if we exclude the Channel Islands. The mean temperature in summer is 58°, and in winter 45°. The prevalent wind is south-west, or west-south-west. As a consequence of this equable temperature, there are numerous plants growing in the open air at Tresco, in the garden of Mr Smith, the lord of the isles, which at Kew are to be seen only in the hothouses. The aloes are magnificent : and rare plants from California and New Zealand flourish in profusion. From this you perceive that the climatal conditions are very favourable, and, whenever Rennel's Current permits it, the southern forms of animal life are swept in by the Atlantic.

One great condition demanded by the tide-hunter is wanting here. There are no caves, no gullies, no huge dark fissures, few overhanging ledges and rock-pools. It has already been noticed in these pages that darkness and depth seem primary conditions even on a good coast. Within the sheltering darkness of caves and fissures, these animals, impatient of the light, are to be found crowding together, and are only accidentally found elsewhere. On such a coast as this you gain nothing, unless you have a man and crowbar with you to turn over the big stones. Under these stones the animals crawl and nestle, chuckling, no doubt, at your zoological despair in the helpless endeavour to get at them, but laughing on the other side of their mouths (by a remarkable anatomical mechanism not yet explained) when they find that you have outgeneraled them, and have overturned their bastions. And yet this love of darkness is very paradoxical. Some of them, especially Annelids, are so impatient of the light that they speedily die in your jars and bottles, unless abundant shadows protect them. The Actiniæ are stimulated by the light; or perhaps it would be more accurate to say that the effect of light upon the sea-weed oxygenates the water, and thus makes the Actiniæ more vivacious. Some Actiniæ—the *Daisies*, for example—habitually flaunt in the exposed glare of sunlight; but the majority, like all worms, Crustacea, and most Molluscs, retire into the darkest shade they can find.

This has a paradoxical air, when we reflect on the paramount importance of light among vital agencies. In darkness the infusoria, it is said, will not develop. In darkness the plant withers. Try to rear a plant in darkness, and no amount of heat, air, or moisture (the other vital agencies) will stimulate it to the processes of real growth. Deprived of light, it is deprived of food, and the possibility of food. It then lives entirely on its own substance, like a starving animal; the store of food which was in the seed is used up, but no new food can be assimilated from without. Nay, if the experiment be carefully conducted, you will find that your plant in darkness, in spite of *apparent* growth, has really lost some of its substance, instead of increasing it; weighing less, when dried, than the dry seed from which it issued. Science has proved that it is in light, and in light alone, that plants deoxidise carbonic acid — setting free the oxygen, which can then be breathed by animals, and in thus setting free the oxygen, releasing the carbon, which nourishes the tissues of the plant. It was thought (and is still printed in many text-books) that the green parts caused the liberation of oxygen in light; but Mulder corrects this, saying that the parts do not liberate the oxygen because they are green, but become green in the process.[*] Rear the plant in darkness, and its leaves will be pale; bring it into sunlight, and these pale leaves instantly decompose carbonic acid, and assume a green tint.

The history of our knowledge of the relation between light and organisation is soon told. It was not suspected until 1771, when Priestley discovered that the plant gave out an air which was capable of maintaining combustion. He allowed a burning candle to extinguish itself in a closed vessel, into which he subsequently introduced a living plant; and in ten days this plant had so altered the condition of the contained air, that the candle once more ignited in it. Many

---

[*] MULDER: *Versuch einer Physiol. Chemie*, 1851. A translation of this valuable work was published under the auspices of the late Professor Johnston.

a schoolboy can now explain this, which was then a splendid discovery, and opened the path whereon, three years later, Priestley laid a foundation-stone of modern chemistry. Progress in science is at times unaccountably slow. For fifteen years had Europe been acquainted with the fact that growing plants set free "oxygen," as we call it; but no further step was taken, till Ingenhouss showed that this oxygen could only be developed by plants *when in sunlight*. Neither he, nor any one else, suspected whence came this oxygen; that was a mystery for another ten years, when Sennebier's work[*] gave to science the simple and pregnant fact, that sunlight enables the leaf to liberate oxygen from the carbonic acid of the air. He proved that sun-heat alone would not suffice; sunlight was the agent at work. Living physiologists have even separated the particular ray of sunlight which exerts the intensest effect. Professor Draper was the first to show this. In his recent work he says: " Since the sunlight is composed of many differently coloured rays, and different principles, it becomes an interesting inquiry which of these is the immediate agent in ministering to the nutrition of plants. In 1843, by causing plants to effect the decomposition of carbonic acid in the prismatic spectrum, I found that the yellow is by far the most effective, the relative power of the various colours being as follows: yellow, green, orange, red, blue, indigo, violet. My experiments on the production of hydrochloric acid

---

[*] SENNEBIER: *Sur l'Influence de la Lumière solaire pour metamorphoser l'Air fixe en Air pur par la Végétation.* 1783.

by the direct union of chlorine and hydrogen under the influence of light, both solar and artificial, conclusively establish the fact that the primary condition essential for the chemical action of light is the absorption of some particular ray. If the physical condition of substances otherwise easily decomposable is such that they transmit light without absorbing any, no chemical change ensues in them, and the same in cases of combination. Thus oxygen and hydrogen cannot be made to unite, even by the most intense radiation, because neither of them exerts any absorptive action; but chlorine and hydrogen unite with energy, because the chlorine absorbs the indigo ray."*

Such has been the history of this partial withdrawal of the veil which hides the mysterious connection of light with life. And now, reader, as you ramble through the cornfields, and see the shadows running over them, remember that every wandering cloud which floats in the blue deep, retards the vital activity of every plant on which its shadows fall. Look on all flowers, fruits, and leaves as air-woven children of the light.† Learn to look at the sun with other eyes, and not to think of it as remote in space, but nearly and momentarily connected with us and all living things. Astronomy may measure the mighty distance which separates us from that blazing pivot of life; but Biology throws a luminous arch, which spans those millions upon millions of

---

* DRAPER: *Human Physiology:* 1856, p. 461.
† " Blumen, Blätter, Früchte, sind also aus Luft gewebte Kinder des Lichts."—MOLESCHOTT: *Licht and Leben:* 1856, p. 29.

miles, and brings us and the sun together. Far away blazes that great centre of force, from which issues the mystic influence,

"Striking the electric chain wherewith we're darkly bound."

For myriads and myriads of years has this radiation of force gone on ; and now stored-up force lies quiescent in coalfields of vast extent, which was once all pure sunlight, hurrying through the silent air, passing into primeval forests, before man was made, and now lying black, quiet, slumbering, but ready to awaken into blazing activity at the bidding of human skill. From light the coalfields came, to light they return. From light come the prairies and meadow-lands, the heathery moors, the reedy swamps, the solemn forests, and the smiling cornfields, orchards, gardens; all are air-woven children of light.

Not less indispensable is light to animals—first, as furnishing them with plants on which to feed; secondly, as furnishing them with oxygen to breathe; and, thirdly, as stimulating in some unexplained manner the organic processes. Light affects the respiration of animals, just as it affects the respiration of plants. This is novel doctrine, but it is demonstrable. In the daytime we expire more carbonic acid than during the night; a fact long known to physiologists, who explain it as the effect of sleep; but the difference is mainly owing to the presence or absence of sunlight; for sleep, as sleep, *increases*, instead of diminishing, the amount of carbonic acid expired, and a man sleeping will expire more carbonic acid than if he lies quietly awake under the same conditions of light and temperature; so that if less is expired during night than during the day, the reason cannot be sleep, but the absence of light.*

Now, we understand why men are sickly and stunted who live in narrow streets, alleys, and cellars, compared with those who, under similar conditions of poverty and dirt, live in the sunlight. And to give a solid basis to such views, we have Moleschott's striking experiments, which prove that under precisely similar conditions of warmth, age, size, food, &c., the single variation in the condition of light produces an equivalent and constant variation in the amount of carbonic acid expired. In bright sunlight as much as one-fifth more carbonic acid was expired than in feeble light.† And have not all farmers and cattle-breeders unconsciously paid tribute to this principle, by keeping their animals in the dark to fatten them?

Returning from this wide-sweeping excursion to the point from which we started, namely, the love of darkness manifested by our animals, the question arises, How can the paradox be reconciled? One might venture on an hypothesis, were but the facts a little less refractory: when, however, we see one kind of Anemone flaunting in the light, and another creeping under a stone; when we see the Crab impatient of the day, and the Prawn swimming gaily in the brilliant pool; when

---

* MOLESCHOTT, *Licht und Leben*, p. 22, citing the experiments of Böcker.

† See his *Untersuchungen zur Naturlehre des Menschen u. d. Thiere*, i. 12. Also his *Mémoire* in the *Annales des Sciences*, 1856.

we see the Mussel fixing himself by his byssus to the rock exposed to noonday suns, and another bivalve boring his way into that rock, secure from the "garish babbling day;" when, in short, we see no constancy or parallelism in the facts, explanation becomes difficult. Let us be ignorant! Let us acquiesce in mysteries (when we cannot penetrate them), nor vex with noisy questionings the imperturbable reserve of Nature; remembering the words of the poet, that "fools rush in where *gentlemen acquainted with zoology* fear to tread."

For those who enjoy mysteries and paradoxes there can be no lack of such enjoyment here. We walk amid surprises. Only ignorance keeps us from perpetual wonderment; as we lift each corner of the veil, more and more marvellous are the vistas which reveal themselves. My vivarium is as pretty a little world of wonders as a speculative man may need. In this small vase behold two serpent-like fish, with the heads of greyhounds. That fish is named *Syngnathus* by naturalists; "pipe-fish" by less erudite tongues.— (Plate VI., fig. 1.) You see nothing remarkable in it, either as to beauty or eccentricity, and wonder why it has a place among my pets. Listen. When a Basque woman becomes a happy mother, her husband straightway takes to his bed, and lies there in receipt of caudle and congratulations. Mrs Gamp waits on *him;* while the wife pursues her household avocations. To him flock village gossips, copious no doubt in "experiences." He does the "lying-in" with all pomp and circumstance.

Well, our pipe-fish is a Basque in this respect. Strange as it may sound to hear of a fish incubating like a domestic hen, it sounds still funnier to hear that the male fish performs that office, and he alone.

How does he manage it? Remove that gentleman from under the sheltering stone, and you will observe a sort of marsupial pouch formed along his ventral surface, in which lie an immense mass of eggs in two layers of four strands, which completely fill the pouch. Each egg, you observe, is divided into two tolerably equal portions, one half being of a brilliant scarlet, the other opaque white; and occasionally you may observe the scarlet portion divided into two. A pretty sight, is it not? Remove the eggs, and you will find that the pouch is a mere fold of the integument, and that the eggs are as much *outside* the body as those of a hen are, in the nest. So that the male fish does veritably incubate; and I hope the *Syngnathus* from this day forward will have an interest for you.

Indeed, the fish-world presents us with many anomalies, which press heavily on our generalisations, and make us relinquish them one by one. As a sample, let us consider this plausible passage, wherein maternal emotions are constructed out of animal heat: "Still more remarkable is the effect of a mere exaltation of animal heat upon the instincts and affections of the different races of the Vertebrata. The fishes absolutely unable to assist in the maturation of their offspring, are content to cast their spawn into the water, and remain utterly careless of the progeny to be derived from it. The reptile, equally incapable of appreciating the pleasures connected with maternal care, is content to leave her eggs exposed to the genial warmth of the sun until the included young escape. But no sooner does the vital heat of the parent become sufficient for the purpose designed by Nature, than all the sympathies of parental fondness become developed."* This is a very plausible generalisation; but there are facts which peremptorily contradict it. On the one hand, there are cold-blooded vertebrates—fishes, such as the Hassar, Goramy, Stickleback, Lepidogaster, and Syngnathus—which make nests, or sit on their eggs. On the other hand, there are warm-blooded vertebrates—birds, such as the cuckoo and American cowbird, which, utterly regardless of maternal delights, leave their eggs to be hatched by other birds.

The fishes contradict our generalisations on many other topics; and a very curious passage in Natural History might be written by any one who should take the trouble to collect and group together what may be called fish-paradoxes. Thus there are fish that fly; fish that climb (*Percha scandens*); fish that hop, like frogs, using their fins as veritable legs (*Lophius*); fish that ruminate (*the carp*); fish that discharge electricity in sufficient intensity to decompose water; fish that migrate; fish that make nests; fish that incubate; and fish that bring forth their young alive.

Fish that sing have not yet been heard, but that some of them make an approach to vocal performances

* RYMER JONES, *General Outline of the Animal Kingdom*, p. 615.

by emitting tones, was known to Aristotle, who specifies six different kinds; and Johannes von Müller has recently collected the literature of this subject in an interesting essay,* in which, after giving his own observations, he explains the mechanism by which the sounds are produced.

To these, recent researches have added facts even more amazing to the systematic mind, namely, that there are fish which *normally* are double-sexed; and at least one species which undergoes *metamorphoses* similar to the metamorphoses of reptiles.†

But we must not linger over the fish, when so many other animals call for notice. The Actiniæ distributed among these vases and pie-dishes will convey some idea of the wealth of Scilly in such creatures. Here are *Gems* and *Daisies*, *Antheas* and the lovely "orange-disked," by Gosse named *Venusta*. The *Crassicornis*, you observe, is represented in every variety of splendour. Here is one with a rich green body and white tentacles; here another with dark-red body and buff tentacles; a fourth presents his scarlet beauty to our gaze; a fifth is ravishing with carnation tentacles barred with white. Here is a tiny *Actinia nivea*. Here are three of a species new to me. They stand an inch and a half in height, with a tendency to elongate themselves still further. They have but one row

---

* *Müller's Archiv für Anat. u. Phys.*: 1857, p. 249.

† For the first of these, see the researches of M. Dufossé in the *Annales des Sciences Naturelles*, 1857; for the metamorphosis of the Ammocete into the Lamprey, see *Müller's Archiv für Anat. u. Phys.*: 1856, p. 323.

of tentacles, not retractile; and their white bodies are encircled with rows of reddish spots, some small, others much larger, the latter surrounded with a ring of white. The colour of their tentacles is dark green, spotted with brown. They most resemble the Anemone which is found, I believe, in Weymouth Bay, of which I have one exquisite specimen—translucent white spotted with red, the spots crowding towards the base; the tentacles pure white, with a brilliant apple-green streak running down on either side, and passing over the oral disc to the delicate pink mouth. Here is another novelty, in size about one-fourth of an inch in diameter, the body delicate French grey with white strips, and tentacles of pure white. And here is that lovely Lamp-Polype, *Lucernaria*, with its little knobbed tentacles contracting and expanding.

Let us pass with a mere glance at those Eolids, old acquaintances, and at that solitary Ascidian, passing his existence in the somewhat monotonous opening and shutting of his two orifices (the only visible sign of life he gives), to pause for a moment over the Echinoderms. There a *Goniaster* is clinging to a bit of stone; and there two *Comatulæ* (Plate VI., fig. 2) expand their feathery charms; a single *Sea-urchin* crawls up the side of the dish, and a lovely *Brittle-star* wriggles at the bottom. To look at this brittle-star you would never imagine how sensitively alive he is to insult. Place but a finger on him, and he breaks up his dishonoured body into fragments before your eyes. He thinks no more of throwing away his legs and arms, than a young lord in London thinks of squandering his acres. The late Edward Forbes has left a humorous account of his hopeless endeavours to secure a rare species (*Luidia fragilissima*) in an entire condition. To understand his account, you must know that most marine animals expire immediately on being thrown into fresh water; and you must further be informed that the pigment spec at the end of each arm, or leg, is the extremely hypothetical "eye" of the starfish. Forbes was ready with his bucket, and, "as I expected," he says, "a *Luidia* came up—a most gorgeous specimen. As it does not generally break up before it is raised above the surface of the sea, cautiously and anxiously I sank my bucket to a level with the dredge's mouth, and proceeded in the most gentle manner to introduce *Luidia* to the purer element. Whether the cold water was too much for him, or the sight of the bucket too terrific, I know not, but in a moment he proceeded to dissolve his corporation, and at every mesh of the dredge his fragments were seen escaping. In despair I grasped at the largest, and brought up the extremity of an arm with its terminating eye, the spinous eyelid of which opened and closed with something exceedingly like a wink of derision."

Quitting the Starfishes, let me call attention to those pretty Cowries and the naked Molluscs :—Are not those two *Actæons*, green, with speckles of gold, attractive? I have nothing to tell you about them, however, not having dissected one, nor submitted it to any more rigorous investigation than a casual glance of admiration. Of that magnificent *Doris tuberculata*,

I may have to speak hereafter; meanwhile let us admire the various colours of its cloak, and the delicate beauty of its frilled branchiæ, for there is nothing in its general demeanour to admire. It has no pretty ways to captivate our hearts—a mere drawing-room beauty, large, lazy, lymphatic, and unintellectual. This other *Doris* has not even brilliant colours to attract notice: a dirty white cloak is thrown over its person, which, except the delicate gill-tuft, has really nothing to boast of. But as Falstaff consoled himself with the thought that his ragged troop were "mortal men, food for powder," and as good for bullets as a troop of better men, so I estimate this *Doris* with an anatomical eye, and find it worth attention. The *Eolids* are poorly represented here—only two, *E. Papillosa* and one *E. Alba*; but there happen to be abundant specimens of the *Pleurobranchus* (see Plate VII., fig. 3), a naked mollusc of translucent buff colour, which on the rocks at first I mistook for a *Doris*, but found on inspection could not be one; and recourse to Woodward's *Mollusca*, and Gosse's *Handbook*, at length satisfied me of its title and position. This animal wears his gill drooping from his side, under the cloak, with the jauntiness of an ostrich feather drooping from the side of a lady's hat; and instead of carrying his shell like a breastplate or backplate, he wears it beneath his skin, as timorous tyrants used to wear mail beneath their clothes. The *Pleurobranchus* was a novelty to me, and when the fisherman who accompanied me, to turn over the stones, first pushed aside the stone under which it crawled, I expressed my enthusiasm by at once promising him an enlarged fee—a most impolitic action on my part, and one which completely unsettled my companion's mind. From that moment he became a bore. Every animal I condescended to bag, became the object of his loudest laudations, in the dim hope that somehow he might persuade me I had secured a brilliant specimen, one causing fresh overflows of generosity on my part. "Well, he be a beauty! We arn't seen one like him before, I reckon? He's worth a sovereign, I'll bet a guinea!" This was the running accompaniment he kept up, as he handed me an Anemone or a bit of Sponge. The Sponges especially alternately excited and damped his hopes. He was constantly exclaiming "Oh! look here, then! what be this?" and as constantly hearing, "Only a Sponge, Pat," which greatly moderated his ardour. One moment I thought he was going to persuade me the Sponge was immensely valuable, but he digressed into safer admiration of the Annelids just captured. In fact, as I said, my outburst had been most impolitic, by rousing visions of El Dorado. From that moment his conversation pointed with fatiguing monotony in the one direction of extra fees. The next day I took another man, and we found more specimens of the *Pleurobranchus* than I had room for. A dozen were brought home; and as—to judge from all the works accessible in Scilly—the anatomy of this mollusc has not been studied since Meckel described it, these dozen specimens will afford me ample means of investigation: meanwhile, the account given by Professor Owen of the digestive organs is sufficiently curious to be quoted. The animal has four separate stomachs: " The first, which is membranous, receives the bile by a large orifice placed near its connection with the second digestive cavity, which is smaller and more muscular; to this succeeds a third, the sides of which are gathered into broad longitudinal lamellæ, precisely similar to those of a ruminant; and to render the analogy still more perfect, a groove is found running along the walls of the second cavity from one orifice to the other, apparently subservient to rumination. The fourth stomach is thin, and its walls smooth."* A mollusc equipped with the ruminating series of stomachs, is paradoxical enough; but what shall we say to this ruminating Mollusc when we find him *not* to be a vegetable feeder?

One of these Pleurobranchi is resting on a whelk-shell inhabited by our friend the Hermit-crab, of whose habits we learned something at p. 49. I did not then know that the fine old naturalist Swammerdamm had argued, and very ingeniously too, against the belief that this crab inhabited the shell of another animal, which he calls a fable, and which on anatomical grounds he endeavours to disprove, declaring our friend to be a Crab-Snail.† Although he was unquestionably wrong in this, he was right enough in laughing at Aristotle and Ælian for asserting that the crab lived *with* the mol-

---

* Owen: *Lectures on Comparative Anatomy*, p. 558.
† Swammerdamm: *Bibel der Natur*, p. 64, *et seq.*

lusc, and undertook the office of finding food for both.

Passing from the region of vases and pie-dishes, let us enter that of wide-mouthed bottles, not so attractive to the unlearned eye, but full of promise to the mind which sees there Polypes, Polyzoa, and ova. For these we want the microscope, one of those "intellectual tubes which give thee a glance of things that visive organs reach not;"* and many a blissful hour may we spend over its revelations. We may hear, indeed, that our perplexed vassal reports us as spending the day "a-squinting through a glass;" but her sarcasm is harmless, and the revelations are thrilling. What can be more interesting than to watch the beginnings of Life, to trace the gradual evolution of an animal from a mass of cells, each stage in the evolution presenting not only its own characteristics, but those marks of affinity with other animals which make the whole world kin? To watch the formation of blood-vessels, to see the heart first begin its tremulous pulsations, to note how Life is, from the first, one incessant struggle and progress —these keep us with fascinated pertinacity at our study.

Among other things, I have watched the development of the *Eolis* and *Doris* with great interest; not the less so from the fact that, in spite of the marked differences between the developed animals, their course of development is so indistinguishably similar. On the rocks, or on the side of your vases, you may see a long coil of spawn, looking like delicate pearl beads enveloped

* SIR THOMAS BROWNE.

in a perfectly transparent membrane. The first thing which will surprise you, on commencing the investigation, is that the division of the yolk-mass is unlike that of most other eggs. In the first place, it is not symmetrical; in the next place, it is not always the same. Sometimes the division occurs in two unequal halves;* sometimes in three, or even four, unequal parts.† I have even counted five. The germ-mass may develop into one, two, three, or even six embryos,‡ all of which are seen slowly rotating in the same envelope; and besides these, there may generally be seen various masses of granules rotating with them, or driven about within the envelope—which are probably fragments of the germ-grass insufficient to form a separate embryo. This multiplication of individuals from one egg, this production of twins, or *triplings*, is a constant fact, and may help the general question of twin births. Very curious it is to watch the increasing activity of the little embryos. At first their rotation is scarcely perceptible; after a while the long cilia protruding from

---

* In the ova of an *Aetaon*, which spawned in my vase, I observed the same want of symmetry; the yolk-mass divided in each case into two unequal halves.

† Comp. VOGT: *Annales des Sciences*, 1846. I. 24.

‡ I hesitated to record in the text what I found in my Tenby notebook; namely, that these embryos sometimes amount to as many as six in one chorion, because as the observation was made when I was comparatively new to the subject, and differed from what is said by others, I thought it possible some error of interpretation might have occurred. I have since satisfied myself that my original note was accurate, and I have at this moment a coil of *Aplysia* eggs in process of development, in some of which there are six, seven, and even eight embryos actively rotating in each chorion.

the shell are seen to wave with more vigour, and the animal moves quickly. Just before emerging from its crystal envelope, the rapidity of its motion is very great; and a wondrous spectacle it is to behold many hundreds of them whirling and whirling about till they escape into the water, where they swim to and fro like crowds of tiny Nautili disporting themselves on the ocean.

The mention of Nautili reminds me that these young molluscs, which are without vestige of a shell in their mature stage, are all provided with a good-sized shell in their embryonic stage. According to the principles of Agassiz and others, which would make embryology the principal guide in zoological classification, this transitory presence of the shell would imply that the naked molluscs were higher in organisation than molluscs with shells. This conclusion will not, I think, be accepted. But the fact that the embryo has a shell, of which it is subsequently destitute, is interesting in the speculations it suggests, and will one day, doubtless, receive its due place in science. Curious it is to think of the huge shell of the *Whelk* or *Limpet* fading off into the small shell-plate concealed beneath the skin of the *Sea-hare* and the *Pleurobranchus*, and disappearing altogether from the *Doris* and *Eolis*. Yet perhaps not altogether disappearing; for may not those spiculæ which are so abundant in the integument of the *Doris* represent the shell in a rudimentary condition? I say "represent," meaning thereby that the spiculæ are the analogous product of secretion, not the homologous "skeleton;" for although these spiculæ may stiffen the integument, and in so far fulfil a protective office, I find them in other places—for example, in the membrane which lies next the "brain."

To discover a *new* animal is surely a legitimate pride. We are pleased if among sand-numerous "varieties" we can alight upon even a new variety, and affix our names to it; but a new animal—something no prying zoologist has ever seen the like of before, something no "plodding German" has described, something we can call our own, and having given it a Greek name, write with modest glory *mihi* after it, instead of Linnæus, Cuvier or Owen—is not *that* a pleasure and a pride? But you must be very circumspect, or you will find, as I did, after long examination and some parental pride, that some "plodding German" *has* been before you. One day looking down upon a tuft of red sea-weed (*Polysiphonia*), on which were clustered several specimens of *Lagenella repens* (one of the Ciliobrachiate Polyzoa, so thoroughly investigated by Dr Arthur Farre and Van Beneden),* I observed a quantity of tiny cups in motion. On removing a bit of weed to the stage of the microscope, I fancied these cups to belong to a new Ascidian; but many examinations gradually dispelled this notion, and left me completely puzzled. I ransacked my books in the vain effort of identification, and began to think the animal before me was a novelty.

---

* FARRE: *Observations on the Minute Structure of some of the higher forms of Polypi.* (Philosophical Transactions, Part II., for 1837.)

VAN BENEDEN: *Histoire Naturelle des Polypes Composés d'eau douce.* 1850.

This suspicion grew into a conviction; and, after bestowing a proper Greek name on it, I made several preparations to show admiring friends. The animal springs from a creeping stem, and stands about the tenth of an inch in height, or less. Each individual is connected with every other by this creeping stem, and consists of a vase-shaped body, or cup, supported on a stalk. When the animal is fully expanded, it unrolls the edge of its cup into a circle of twelve or fourteen ciliated tentacles, curled downwards like the young fern-leaf, or like the handle of a Greek vase. These tentacles are, as I said, cut out of the edge of the cup, not enclosed in the cup, like those of a polype. The alimentary canal is a long convoluted tube; the cavity is lined with cilia, and at the bottom there is a mass of yellowish granules (hepatic cells?) and occasionally the food may be observed rotating, as if on an axis.

Some weeks after convincing myself the animal was new, I dredged between the coasts of Jersey and Brittany a small *Pecten*, on the shell of which, besides other animals, there was my new friend in great activity, and of much larger size than the one got at Scilly. It was then I learned that my friend had already been described—that, in short, it was the *Pedicellina* of Sars, or at any rate differed therefrom only in such unimportant particulars (such as the retractility of the tentacles) as would at the most constitute a distinction of species. I made this out by studying the development of the animal. In Owen's Lectures there is a diagram of the embryonic phases of *Pedicellina*, and some of these were what I had already drawn from my own animal. One fact, however, is worth mentioning, because, as far as I can ascertain, it is not known; namely, that the *Pedicellina* is *viviparous*, as well as oviparous and gemmiparous. While examining a cluster of them, I saw something protruding from the mouth; presently another something rose beside it. I watched anxiously, with a certain flutter. I suspected they were embryos. Slowly they emerged, and the suspicion grew stronger and stronger, till finally three ciliated embryos, in the stages of development indicated at figs. 7 and 8 in the diagram,* swam away in the water. There could no longer be any doubt that my Scilly animal was a species of *Pedicellina;* but I had the compensation of having found, instead of a new animal, a new fact with respect to its generation.

This has been narrated as an illustration of the caution necessary before announcing new genera and species to the world, and needlessly encumbering the already unrememberable lists of names. I was also interested by the puzzlement into which I was thrown as to the classification of my new animal (when it was thought to be new).

Indeed, the assignment of animals to their proper places in systematic classification will continue to be the work of much unsuccessful ingenuity, until more rigorous and philosophical principles of classification be adopted. That present classifications are only pro-

---

* Owen: *Lectures on Comparative Anatomy*, p. 152.

visional, will scarcely be denied. They have not the stable basis which can make future researches the simple extension and application of existing principles. A new method is inevitable; but we may be years before it is promulgated. An instructive example of our inability to apply the present Method, otherwise than in a provisional way, is afforded by that puzzle to zoologists, the *Sagitta bipunctata* (Plate V., fig. 1.) Nobody knows where to place it. In aspect it is fish-like; in some structural peculiarities it is fish-like; in others it is molluscan; in more it is annulose. Siebold classes it with mollusca; Huxley and Krohn, with the annulosa, the former pointing out that "it presents equally strong affinities with the four principal groups—1. The Nematoid worms; 2, The Annelida; 3. The Lernæan Crustacea; 4. The Arachnida."*

Place it where we will, the animal is very interesting, either when darting about in a glass vase flapping the water with its tail, and fixing itself to the side of the glass (using the vent as a sucker?), or seen on the microscope stage, where its extraordinary transparency obliges a liberal use of "stops." It is then seen to have a head with a formidable set of hooks (which, however, do *not* seem to fulfil the office of jaws), and two large eyes. The narrow body is divided into two equal lengths (in my specimens this was so; in the figures published by Mr Busk and Mr Gosse the anterior portion is considerably the larger); in the upper half lies the straight alimentary canal, terminating in

* *Report of British Association*, 1851. Sections, p. 78.

a ciliated orifice; on either side of the canal lie the ova; in the lower half, which is longitudinally divided by a septum, the whole cavities are filled with granules of various sizes, moving, by a scarcely perceptible progress, round and round, like food in the stomach; and these granules prove to be the spermatozoa which issue from the two orifices near the caudal expansion. My species, which I have christened *Sagitta Mariana*, is about the quarter of an inch in length, and differs in several points from the species described and figured by Mr Gosse in his *Tenby* and *Handbook*, and by Mr Busk in the *Microscopical Journal*.* For instance, it has no anterior fins; and the posterior fins, which arise near the oviducts, are continuous with the caudal fin, into which they expand; that is to say, it has really but one fin on each side, with a caudal expansion. Another peculiarity worthy of notice is, that, in consequence of this union of the lateral and caudal fins, the orifices through which issue the spermatozoa, instead of opening directly in the integument of the body, are openings in the fin itself; as I have convinced myself by repeated examination—a circumstance which leads us to suspect that the "fin" is only a membranous expansion of the integument, and not properly a fin. Other details, not mentioned, and therefore, I presume, not present in the specimens previously described, but which were constant in those found at Scilly, are the double band of light yellow

---

* October 1855. In this paper the reader will find a summary of all that was then known on the subject.

granules forming three sides of a parallelogram about the œsophagus, and two dark-brown irregular masses above each oviduct. The hairs, or spines, are distributed over the fins as well as over the body—an arrangement which has been noticed by Khron, who denies that they are *setæ* at all, considering them to be merely epidermic processes. On what grounds he so considers them, I am not informed; but I entirely concur with him, because I find these supposed *setæ* very rapidly undergo decomposition, which would not be the case were they inorganic hairs or spiculæ. Let me conclude these perhaps dry details with the remark that the delicate layer of epidermis, composed of rounded cells—the existence of which Krohn first disputed and then admitted—was visible in my specimens, although I mistook it for scales or scaly epithelium; and that I can confirm Huxley's statement of the existence of a ciliated oviduct in the external part of the ovary, if the statement of so accomplished a zootomist needs confirmation.

But the point of all others which in this interesting *Sagitta* excited my interest was one I have not seen noticed by others, namely, the entire *absence of any vascular system*. Here is an animal with a nervous system of some importance (see Prof. Huxley's diagram in the *Microscopical Journal*) with eyes, if no other organs of sense, and with muscular fibres of the *striped* kind; yet in spite of such characteristics of high organisation, it is totally without "blood," without a trace of a vascular apparatus. So striking a paradox necessarily fixed my thoughts for some time, till at length light seemed to break in obliquely from some investigations pursued respecting the relation of the blood to Respiration. These investigations are not yet sufficiently advanced for publication, but they point unequivocally to the fact, that in the animal series there is a *definite relation existing between the development of the vascular and respiratory systems,* the specialisation of the one following the specialisation of the other.* Seen by this light, the *Sagitta* ceases to be paradoxical; its respiration is performed by the whole surface, without the need of any special organ such as gill or lungs, and this absence of a respiratory apparatus carries away with it the need of a vascular apparatus. No Respiration, no Circulation: the one necessity creates the other.

If the *Sagitta* is without a vascular system, it must consequently be without blood. Even so. Without "blood" it is, unless we extend the term "blood" to every fluid fulfilling the office of a nutrient fluid; an extension not only obliterating the whole purport of exact language in science, but finally reducing us to the state of the Irishman who saw in a lake "all the materials for punch—barring the whisky and the sugar and the lemons;" since when we descend to the simplest forms of organisation, we reach a nutritive fluid which is water, and nothing more. Dr Thomas Williams, to whose researches on the blood we owe

---

* Compare on this point BERGMANN u. LEUCKART: *Vergleichende Anatomie,* p. 170.

grateful acknowledgment, considers that in the Echinodermata the blood-proper first makes its appearance; below that point there is no blood, but only chylaqueous fluid; and even for several stages higher, this chylaqueous fluid continues to hold its place beside the true blood; so that a worm, for instance, has two fluids —blood, circulating in a system of closed vessels, and chylaqueous fluid oscillating in the general cavity outside the vessels.*

It is indispensable, in philosophic zoology, to discriminate between *blood*, a fluid of definite constitution *circulating* in a system of vessels, and the *chylaqueous fluid*, formed of water and the products of digestion, *oscillating* in the general cavity. But my investigations lead to a still further reduction of this latter fluid; and instead of saying, with Dr Williams, that the simplest condition of a nutrive fluid is a "very dilute solution of albumen in sea water,"† I am forced by facts to say that lower even than this is the earliest state of the nutritive fluid: namely, sea water carrying certain gases and organic particles, but *without* definite chyle-corpuscles, such as Dr Williams figures—without even albumen in solution, at least as a *constant* element. This is the case with all the Sponges. They simply suck in sea water and expel it. The reader will, however, learn with surprise that this also is the case with the far more highly organised *Actiniæ;* a

---

* See page 71.

† See his paper on "The Blood" in the *Brit. and For. Med. Rev.*, Oct. 1853, p. 480.

fact which, when coupled with what was said in the preceding chapter respecting the non-digestive powers of these animals, may lead to many interesting speculations.

If you have ever kept Sea-Anemones, or have even paid casual attention to them in the vases of your friends, you must have noticed their remarkable variations in size. The *Crassicornis* which excited your cupidity by his magnificent proportions, as the eye first beheld him in the rock-pool, has collapsed to a fourth of the size before you have chiselled him off; and in collapsing he squirted continuous streams of water from his pores and tentacle-tips.* That *Gem*, which an hour ago was expanded to the height of an inch, is now a mere button. The ordinary explanation of this phenomenon is that the animal swells itself with water, which it violently ejects on being "irritated" or "alarmed." But as we are just now looking with scientific seriousness at our animals, we will discard all anthropomorphic interpretations, like those which point to "alarm," because they not only confuse the question, but lead to awkward issues; among others, that the Anemones have highly susceptible souls, as liable to emotions of alarm as a fine lady. When they are in undisturbed quiescence in pool or tank, the same ejection of the water takes place, only with less rapidity. Their normal condition is that of

---

* Those anatomists who still deny the existence of openings at the tips of the tentacles, need only "irritate" a *Crassicornis* to be convinced of the fact.

constantly sucking in and pouring out sea water, *for on this mainly depends their nutrition.* Keep one in a vase, without feeding it, without even suffering visible food to float in the water, and it will nevertheless feed and flourish simply by this absorption of water, which contains gases and invisible organic particles.

I had three *Daisies* in a vase for nine weeks, during which time they were entirely without food, except such as the water held in solution. They were as healthy and active all the time as any others, and I believed they would have continued so to the end of the chapter. The experiment is worth trying.

Far as this is from the notions current respecting the nutrition of Anemones, it is easily demonstrable. In the preceding chapter I showed that the supposed " digestion " of the Anemone was confined to the pressing out of the juices, and the rendering soluble, by *maceration,* of organic substance; I have now to show that this animal is not only without " blood," in any proper sense of the term, but also without that simpler form of blood named " chylaqueous fluid " by Dr Williams and succeeding writers.

This will probably startle the reader, especially if he happen to have seen the writings of Dr Williams, who actually figures the chyle-corpuscles of the Actiniæ, and declares that the fluid gives an albuminous reaction. But with the highest respect for that observer, repeated investigations, which have subsequently been confirmed by the well-known zoologist, Mr R. Q. Couch, compel me to declare that *no* such fluid circulates in the Actiniæ—an assertion which can readily be tested. The water is easily forced out of the tentacles, or collected by cutting open the Actiniæ in a glass. Evaporate it, and you will find it to be sea water, holding sometimes organic particles in solution. Test it with concentrated nitric acid, and instead of becoming turbid, as it would if it contained albumen in solution, it remains unaltered, except that, when organic particles are present, they become distinct. Examine the fluid with the microscope, and you will find animalculæ and various particles, but nothing like definite corpuscles, such as are visible in the true chylaqueous fluid. It is, in short, sea water and nothing more.

Feeling that in thus opposing the positive statements of an accomplished zoologist like Dr Williams, I might very possibly be under some error of observation, or interpretation, I requested Mr R. Q. Couch to repeat the investigation, that I might either be corrected or confirmed. He very kindly undertook the task, and thus wrote: "I took two specimens of the *spotted Mesembryanthemum*, and forced the water from the tentacula, and found, under the microscope of 300 linear, numerous infusorial creatures rapidly moving about. On treating this with nitric acid, I had a slight opalescent deposit, or rather a diffused milky cloud of very slight character. The next day I obtained several from our rocks, and again forcing the water out, I obtained two specimens of a microscopic nudibranchiate mollusc, and other creatures of similar character."

I may observe that the slight milky cloud here spoken of occurred once, and only once, in my examinations, showing it therefore to arise from an accidental, not a *constant* element. "A third and a fourth experiment," continues Mr Couch, " was made on others taken from our rocks in a contracted state, and consequently empty. I placed them in sea water, which I had repeatedly *filtered* through sand, and afterwards cloth. In this they remained till to-day, when, taking them in an expanded state, I put them to drain in a small glass dish. *In this I could discover nothing organic, and it gave no cloudiness by nitric acid.* As these experiments are quite in accordance with others made some years since as regards their results, I regard this fluid as merely sea water free from every admixture of secreted matter."

Nothing can be more explicit, however startling the result. In the presence of such evidence, one is amazed to re-read Dr Williams when he says: "The surrounding water enters the stomach, where it briefly sojourns, then passes through the opening at the bottom of the great cavity of the body: in this cavity it remains for a variable period; it now injects the tentacles. *Corpuscles now arise in the fluid;* it becomes thicker in consistence through increase of albumen: it is no longer pure lifeless sea-water; it is a corpusculated chylaqueous fluid: it is competent to serve the ends of nutrition. Whence do the floating cells proceed?—what produces them? Certainly no solid organ; neither liver nor spleen can in this case interpose its agency. Then, is it possible that there can inhere in albumen a mysterious histomorphotic power in virtue of which it transmutes itself from the liquid into the solid condition? This were only a mode of enouncing the theory of spontaneous generation."* All these questions are superfluous, since the fact is imaginary; an albuminous corpusculated fluid does *not* circulate in the cavity of the Actiniæ; sea water, carrying whatever accident may have brought to it, is the "nutritive fluid" of these animals.

Dr Williams has, however, published drawings of the corpuscles discoverable in the fluid, and Schmarda, as I learn from Victor Carus,† declares that such corpuscles are constant. Can these statements be reconciled with what results from the experiments of Mr Couch and myself? They have the advantage of being positives against negatives, and must, one would think, have some truth in them. What is that truth? This question, I confess, haunted me till an answer suggested itself. One of my *Daisies* (*A. Bellis*) brought forth a round mass of fifteen young, agglomerated together in a ball: they were in different stages of development, and being perfectly transparent, admitted of easy microscopic examination. In them spherical globules were distinctly visible, circulating by the action of the cilia lining the cavity; and with the globules an occasional animalcule. This was the case with all of them; and

---

* *Loc. cit.* p. 483.

† *Jahresbericht über die im Gebiete der Zootomie erschienenen Arbeiten*, 1856, p. 25.

as the globules seemed most abundant in the youngest specimens, the idea occurred to me that Dr Williams had only examined the fluid of young *Actiniæ*, and had concluded that what was true of them was true of adults. This idea, however, grew less and less plausible on reflection. That a young animal should have a circulating fluid higher in character than the fluid of the adult, seemed unreasonable. While thus speculating, I observed a great irregularity in the size of the globules; sometimes they seemed united together in considerable masses. To pursue the investigation closer, I opened the cavity with a needle, and let out the fluid: to my surprise, these floating globules turned out to be no chyle-corpuscles, but the yellow spherical cells (?) abundant in the tentacles of the adult *Daisy*, and the tentacles of the *Anthea*, and which give the brownish colouring to its body (see p. 165). What may be the function of these yellow spheres, I know not; but it is certain they are not the corpuscles of a circulating fluid (they are stationary in the adult), although I must suppose Dr Williams and Professor Schmarda have mistaken them for such, since no other definite globules are discoverable; and these circulate only in the young.*

Parenthetically it may be mentioned, that in Professor Allmann's recently published Monograph on the *Fresh-water Polyzoa*, the subject of the chylaqueous

---

* In the *Annals of Natural History*, 1858, vol. i. p. 174, Mr GOSSE has published observations in support of Dr WILLIAMS's opinion, but his observations seem to me wanting in the requisite fulness and rigour.

fluid is thus noticed: "It is by no means homogeneous, and numerous corpuscles of very various and irregular shape may be observed to float through it and be carried about by its current. Some of these corpuscles are, doubtless, spermatozoa; others are of no definite shape, and look like minute portions of the tissues separated by laceration. If it be admitted, as I think it must be, that the perigastric fluid consists mainly of water which has obtained entrance from without, it then corresponds to a true aquiferous system subservient to a respiratory function. But it also, without doubt, receives certain products of digestion, which had transuded through the walls of the alimentary canal; it thus connects itself with the digestive system. It is, moreover, the only representative of a sanguiferous circulation, for in the Polyzoa there is certainly no trace of a heart, nor can anything referable to a true vascular system be detected."[*]

Returning to the fluid in the cavity of the Anemones, we see the necessity of a cultivated caution in the acceptance of statements in matters so complex as those of Biology. The respect justly due to Dr Williams as an investigator, has caused his views respecting the "blood series" to be accepted without verification. As I make it a rule to verify every one's statements, when they fall within my own investigations—believing with Harvey "how unsafe and degenerate a thing it is to be tutored by other men's commentaries, without making

---

[*] ALLMANN: *Monograph of the Fresh-water Polyzoa*. Published by the Ray Society, p. 23.

tryal of the things themselves, especially since Nature's Book is so open and legible"*—I determined to do so with respect to this on the fluid of the Actiniæ. The result has been seen. It throws a new difficulty in the way of rightly understanding the processes of Nutrition; but it is a step towards a right understanding, because it removes an explanation which, seemingly true, masked the real process. It also gives the final blow to those gratuitous determinations of special "organs of secretion" in the Actiniæ in which zoologists have revelled. If there is no blood, there can be no secretions from the blood; and all attempts at fastening a secreting function on the "convoluted bands" may *a priori* be dismissed: I say *a priori*, because no one has yet attempted, by chemical tests, to prove the presence of bile, or urea, or any other product of secretion, in these organs; and as, therefore, the function is assigned on *a priori* grounds, on those grounds may it be dismissed.

On closer inspection this conclusion becomes more imperative. The enormous mass of these convoluted bands, forming by far the largest organ in the body, forbids the idea of its function being that of secreting urea. It is true that Bergmann and Leuckart suggest this to be their function; so does M. Hollard; and Victor Carus speaks with some decision on the point.† But I would ask these eminent writers how they reconcile such a supposition with anatomical and physio-

---

* HARVEY: *Exercitations concerning the Generation of Living Creatures.* 1653.

† V. CARUS: *System der thier. Morphologie*, p. 148.

logical considerations? There are no vessels in the Actiniæ which would convey the products of disintegrated tissue to these secreting organs. If such products are thrown into the fluid of the general cavity, they will be got rid of as that fluid passes out of the body, without undergoing a preliminary secretion in the cells of these convoluted bands. Seeing the disposition of these bands, attached to the membrane called "Mesentery" (Plate III., figs. 2 and 3), on one side, and on the other floating free in the cavity, we detect no means by which the disintegrated products, urea, &c., could reach them. Moreover, the first step in organology must be to determine whether the *product* of the organ is present, *e. g.* ova in ovaries, bile in liver, urea in kidneys, and so on; and until chemical reagents have detected urea or bile in these convoluted bands, we may rest on the assurance that these bands are neither urinary nor biliary organs. To look for such special organs in so simple an organism, seems to me like seeking for a circulating library in an Esquimaux village.

The mention of a library carries my thoughts, by an easy transition, to our evening studies. When the labours of the day are over, the microscope is put up, the work-table is quitted, and the delicious calm of candle-light invites us to quiet intercourse with one of the great spirits of the past, or one of their worthy successors in the present. It is well thus to refresh the mind with Literature. Contact with Nature, and her inexhaustible wealth, is apt to beget an impatience at man's achievements; and there is danger of the mind becoming so immersed in details, so strained to contemplation of the physical glories of the universe, as to forget the higher grandeurs of the soul, the nobler beauties of the moral universe. From this danger we are saved by the thrill of a fine poem, the swelling sympathy with a noble thought, which flood the mind anew with a sense of man's greatness, and the greatness of his aspirations. It is not wise to dwarf Man by comparisons with Nature; only when he grows presumptuous, may we teach him modesty by pointing to her grandeur. At other times it is well to keep before us our high calling and our high estate. Literature, in its finest moods, does this. And when I think of the delight given by every true book to generations after generations, moulding souls and humanising savage impetuosities, exalting hopes and prompting noblest deeds, I vary the poet's phrase, and exclaim—

"An honest book's the noblest work of man!"

PART IV.

JERSEY.

CHAPTER I.

DEPARTURE FROM SCILLY — JERSEY — SCHOOLBOY RECOLLECTIONS — CHOICE OF A SPOT — PREJUDICES AGAINST DISSECTION — ZOOLOGY OF JERSEY — TRAWLING — DIFFICULTIES OF IDENTIFICATION — A STRANGE ASCIDIAN — THE TADPOLE EMBRYO OF A MOLLUSC — A PARADOX — DEVELOPMENT OF MEDUSÆ FROM POLYPES — ALL ANIMALS DO NOT COME FROM EGGS — REPRODUCTION OF THE HYDRA — HISTORY OF PARTHENOGENESIS — A DISCOVERY — ANIMALS PRODUCED FROM GERM-CELLS ONLY — STEENSTRUP'S THEORY — OWEN'S THEORY — A NEW THEORY — THE THREE FORMS OF REPRODUCTION: FISSIPAROUS, GEMMIPAROUS, AND OVIPAROUS — WHAT IS PARTHENOGENESIS?

AFTER seven weeks, the rocks of Scilly appeared to have seen enough of me. A residence so protracted astonished and fatigued them. They knew all my varying moods, and one unvarying, not picturesque costume. Familiar with the ring of my hammer, as it chiselled with savage pertinacity at their granite ribs, they were not less familiar with the compass of my voice, and the extent of my operatic reminiscences, as, seduced by their solitudes, to the orchestral inspiration of their waves I loosened all the power of my lungs in lyrical fervour. For seven weeks had our intimacy lasted, and now there arose the conviction that the time for separation had arrived. Nothing new could possibly be learnt about me. Their curiosity was satisfied, if not satiated; and my presence began to carry a certain monotony with it. Even the two or three meagre dogs, which sniffed about the pier, began to eye me with an air of supercilious weariness; and I forbear to investigate the sentiments of the Scillians, lest they should too painfully resemble the indifference of the dogs. Decidedly it was time to pack up. I took the hint: the Granite Beauties turned a cold boulder on me, and I resolved to weary them no longer. My animals were scattered to the four winds (figuratively, of course—one of the four being the railway to London, which transported a coffee-tin of anemones to a tank-loving lady); my tent was struck, and after hurrying through Penzance, Falmouth, and Plymouth, it was once more pitched in the pretty island of Jersey.

Nothing could be more charming than the welcome smiled by the rich meadow-lands and orchards there. After the bold picturesque solitudes of Scilly, it seemed like once more entering civilised nature. Every inch of ground was cultivated. Cornfields and orchards resplendent with blossoms, sloped down to the very edge of the shore, and, by the prodigality of soil, defied the withering influence of sea-breezes. It was not amazing to me to learn afterwards that the land in the interior yields double the crop, per acre, which can be raised in most parts of England; and that, although the rent is £10 an acre, such rent can be paid by potatoes alone. Elsewhere it is difficult to get even grass to grow close on the shore, and trees have always a look of stunted old-maidenish misery; but here the high tide almost washes the hedge which limits orchards that no right-minded boy could resist robbing. Jersey, indeed, is the very paradise of farmers.

The Americans say that England looks like a large garden. What England is to America, that is Jersey to England. Even the high-roads have the aspect of drives through a gentleman's grounds, rather than of noisy thoroughfares; and the by-roads and lanes are perfect pictures of embowered quiet and green seclusion. There are few more delightful places to ramble in. Every turn opens on some exquisite valley or wooded hill, through the cool shades and glinting lights of which the summer wanderer is tempted to stray, or to recline in the long grass, and languorously listen to the multitudinous music of the birds and insects above and around. Observe, I say nothing of the sea, and the succession of bays on the coast; for what can be said at all commensurate with *that* subject. Even the poets, who not only contrive to say the finest things about Nature, but also teach us how to feel the finest tremors of delight when brought face to face with her, have very imperfectly spoken of the sea. Homer is lauded for having called it "wine-faced." He probably meant some ivy-green potation, since "wine-faced" is the epithet by which Sophocles characterises the ivy.* In any case his epithet is only an epithet, and the sea is of all colours, as it is of all forms and moods. Doubts also may be raised respecting the "giggling" which Æschylus, in a terribly-thumbed passage, attributes to

---

* Œdipus Colon, v. 674, τὸν οἰνῶπ' ἀνέχουσα κισσὸν.

the sea. The " innumerable laughter of the waves of the sea," one is apt to interpret as a giggle—an expression not only unbefitting the sea, but unworthy of the occasion. Neptune was not mocking the agony of Prometheus with a school-girl's incontinence. He was too grand and fluent for such weakness. In moments of serenest summer-calm he may be said to smile; in moments of more leaping mirth he may be said to laugh; but to imagine him distorting his countenance by innumerable giggles, would be at all times intolerable, and at *such* a time perfectly indefensible.

On the sea, therefore, allow me to be silent. On the great attractions of Jersey for the naturalist one word will suffice: there is no such spot in England for marine zoology. Besides all these charms, it had other charms in my eyes. Memory consecrated the ground. Eight-and-twenty years ago I was at school here. Changed as the aspect of St Heliers necessarily is, the few spots still recognisable had a peculiar fascination for me. The Royal Square seemed to have shrunk to a third of its old dimensions, but with what strange sensations I first re-entered it! The Theatre had by no means the magical and imposing aspect which it then wore, when it seemed the centre of perfect bliss. Its yellow playbills no longer thrilled me, although memory wandered back to those happy nights when enchanting comedy and tearful tragedy were ushered in by the overtures to " Tancredi," or " Semiramide " (the only two which the orchestra ever played), and when ponderous light comedians in cashmere tights, or powerful tragedians "took the stage" with truly ideal strides. Gone, for ever gone, are those bright credulous days. Never more shall I see the *School for Scandal*, or *Pizarro*, performed as I saw them then. Lady Teazle will never more lure me with her coquettish fan, nor Cora transport me with her drooping ringlets. I can't believe in the vinous gaity and good feeling of Charles Surface; nor think Rolla the most impassioned and eloquent of beings. I know that the sentiments are as unreal as the acting, or the stage wine and "property" fruit of Charles Surface's banquet. Turning with a retrospective sigh into the Market-Place, I feel the breath of former years rising around me. There is the very corner where we used to "toss" the pieman for epicurean slices of pudding—a vulgar, but seductive form of juvenile gambling. Close by is the spot where we upset "Waddy"—an adipose comrade, much plagued by his leaner contemporaries—flat into an old woman's egg-basket. I see him now, rising covered with the squashed yolks, utterly heedless of the furious imprecations (in unintelligible *patois*), and the furious blows (in perfectly intelligible English) with which the old lady responded; I see his piteous contemplation of his soiled clothes, and hear once more his pathetic exclamation, "Oh damn!" while inextinguishable laughter shakes our leaner sides. Childhood is the Age of Innocence.

Among the changes, it was pleasant to find that no longer did the Pillory disgrace the Royal Square; no longer were criminals publicly whipped through the streets, as I once saw them with shuddering disgust. Formerly women were thus publicly whipped; but that disgraceful exhibition was put a stop to before my time; and now Jersey has grown humanised enough to see that whipping men must be relinquished. It was indeed a loathsome sight. The naked shrieking wretch, with a cord round his neck, halberds pointed at his breast to prevent his hurrying forwards, his back streaming with blood, his face turned imploringly towards the surgeon, who walked beside the executioner, and whom I once heard utter the cruel words, "Harder, Jack!" meaning that the victim had strength to withstand even harder blows—a brutal mob following without sympathy—the procession moving slowly from the Town-Hall to the Prison;—this was the picture Justice frequently presented to the inhabitants of Jersey, and which now, thank God, will never be seen by them again, but will take its place among the brutalities of the past, a sign of the onward progress we have made.

Although St Heliers, "the capital of Jersey," was the spot consecrated by memory, I took up my abode at the entrance of the fishing-village of Gorey, just four miles from St Heliers; and as these pages are addressed to amateur naturalists, some of whom may hereafter visit Jersey, a word on the reason of my choice may not be superfluous. The attractions of the capital I do not deny, and if the visitor is in need of watering-place attractions, he will pitch his tent there; but if his primary desires be zoology and quiet, he will select Gorey, especially during summer, when tide-hunting is necessarily poor, and only by dredging and trawling can he hope to get a good stock of animals. Always go where there are fishermen, that you may have the benefit of their aid. They may bring you what you would never find. It is true there are two sources of difficulty in your way: the first is the almost impossibility of making them understand that you can set any value on things they are accustomed to throw away; the second is, that when you have so tutored them that they know *what* you want, they are strangely backward in their supplies. Money is of course the only cogent argument; yet even money moves them but slowly. They go out day after day, staying out all night, and return often without a shilling's worth of fish; yet although you offer to pay them for oyster-shells and weeds as for fish, they cannot easily be induced to throw this "refuse" of their nets into a bucket, instead of throwing it overboard again. They promise to do so, but you generally wait in vain. At Tenby, in spite of urgent entreaties and liberal promises, only one *Loligo* was brought me; at Scilly, nothing; at Gorey, in spite of my being on the best terms with fishermen whom I had employed, and with whom I had gone trawling, five weeks passed before a bucket of refuse was brought me.* Two words—pertinacity and liberality—sum up the whole art of gaining this desirable result; when

---

* SWAMMERDAMM (*Bibel der Natur*, p. 34) makes the same complaint of the Dutch fishermen, and justly attributes to it the long continuance of our imperfect knowledge of marine marvels.

T

gained, you will need no argument to prove the superiority of a fishing-village.

Comfortably settled at Gorey, and my working-room set in order, I had only to await the spring-tide, once more to gather a variety of pets around me. Not that I was even then without serious occupation. Before leaving Scilly I had put up my Nudibranchs in spirits of wine, and these were now carefully to be dissected. Make no wry face at the word "dissection"—it indicates a very different process from the one you conceive; and as it is one indispensable to the naturalist, I may as well dissipate the prejudice which hangs over it. If prejudices could be satisfactorily displaced by argument, one might ask how a man can pass a butcher's shop with equanimity, yet shudder at the idea of dissecting a rabbit or a dog; but I will admit all such incongruities as facts not assailable by argument, and simply direct the reader's attention to the important differences between dissecting animals of the larger kind, and dissecting our marine pets—it is as great as the difference between knitting a silken purse in a drawing-room, and making a ship's cable in a rope-walk. Almost all our dissections are performed under water, with needles, tweezers, and delicate scissors. There is no visible blood to suggest unpleasant ideas; there is nothing unsightly—to the philosophic eye the sight is full of interest—and if an unsightly aspect be present, has not a noble poetess truly said:—

> "Be, rather, bold, and bear
> To look into the swarthiest face of things

> For God's sake who has made them.
> How is this,
> That men of science, osteologists
> And surgeons, beat some poets, in respect
> For nature—count nought common or unclean,
> Spend raptures upon perfect specimens
> Of indurated veins, distorted joints,
> Or beautiful new cases of curved spine;
> While we, we are shocked at nature's falling off—
> We dare to shrink back from her warts and blains—
> We will not, when she sneezes, look at her,
> Not even to say, 'God bless her.' That's our wrong." *

Nay, has not the greatest of German poets, whose culture of the beautiful was so devout that it has been made a reproach, given us a practical example that not only may Comparative Anatomy reveal its marvels to the delighted eye of a poet, but also that the keen glance of the poet may be that of a great discoverer in anatomy? To Goethe, bones and ligaments were no less beautiful and full of interest than flowers and streams; he saw in them parts of the mystic scaffolding of the temple of life. And laborious and delicate as the amateur may find the dissection of animals to be, he will find his labour well rewarded at the close.

When the spring-tide *did* arrive, it was unfortunately a very poor one; and had Jersey been less wealthy, my hot labours on the rocks would have produced but a meagre result. As it was, I managed to secure an ample supply of *Sea Hares, Eolids, Dorids, Solitary Ascidians, Clavellinæ, Hydractiniæ, Pycnogonida, Actæons, Anemones,* and *Polypes.* In the way of

* *Aurora Leigh.*

novelty there was only the *Hydractinia* (a pretty little white polype growing in clusters on the outside of a whelk shell, inside of which was a hermit-crab) and the *Actinia parasitica*, hitherto only known to me through pictures, but which I found transcending in beauty all power of painting. This beautiful Anemone is extremely abundant here at low tide, but scarcely merits its name of *Parasitica*, for I find it almost as frequently on stones and on the sides of the rocks as on the whelk shells; and in captivity it quits its shell, roaming about the pie-dish, and fixing itself to the side, or to seaweeds, like any other Anemone. The extreme sensitiveness of the *Parasitica* enhances its attractions; it is for ever expanding and retracting its tentacles, elongating, curving, or retracting its stem: sometimes doubling its length, at other times assuming an hour-glass constriction in the middle. The filaments which contain the " thread-capsules" are poured forth in great abundance whenever the animal is disturbed. While on the subject of Jersey Anemones, it may be added that, besides the ordinary species, I dredged what is probably a variety of the *Actinia ornata*, described and beautifully figured by Dr Strethill Wright in the *Edinburgh Philosophical Journal* for July 1856,—the body white, the exterior circle of tentacles orange, the two interior circles white striped with grey, the disc orange in the centre; very charming to behold.

Having stocked my jars and dishes, I was somewhat reluctant to broil in a noonday sun amid the rocks, with little hope of finding any animal not already familiar; and therefore contented myself with the less exciting and more remunerative labour of deep-water hunting. By this I got initiated into the art and mystery of trawling, having made friends with a fisherman, master of a Trawler of about twenty tons.

Pleasant it is on a bright sunny morning, with a nice breeze from the shore, to recline on the deck of a fast-sailing vessel, and listen to the men retailing their experiences, or watch them heave out and haul in the net. Away we glide towards the coast of France, Jersey melting in the distance,—

> "The sands untumbled, the blue waves untost,
> And all is stillness, save the seabird's cry
> And dolphin's leap."

The net is at the bottom, collecting in its gaping mouth the treasures we are duly awaiting; meanwhile, in a sort of dreamy content, we stretch ourselves in the sun till the word is given to haul in, and then anxiety dissipates the luxurious calm. The trawl is a huge net of somewhat conical shape, from twenty to thirty feet wide, from thirty to forty deep. Along the edge of the wide opening is a stout wooden beam. to the ends of which are fastened the trawl heads, namely, thick flat semicircular bands of iron, which serve to keep at a distance of three feet from the beam that portion of the net meant to touch the bottom. In the net there are various pockets. When the trawl is thrown overboard, the weight of the iron carries it to the bottom, the buoyancy of the wooden beam, assisted by the perpendicular support of the iron bands, keeping the upper edge of the net steadily floating three feet above the ground. The rope sweeping along the bottom disturbs the fish; up they dart in foolish distracted haste, and come in contact with the net overarching them; this flurries them, and they dart sideways to escape; in doing which they unsuspectingly swim into the net if they go one way, into the pockets if they go the other. The net, thus scraping the bottom, gathers, of course, a quantity of shells and weeds as well as fish; this is known to naturalists as "trawl refuse," and is always worth careful overhauling.

The contents are all emptied upon the deck, and while Jack is gloating over the turbot, brill, soles, skate, and gurnard, or grimly noticing the utter absence of those desirable individuals, you squat down amid the refuse, and begin a long deliberate investigation thereof. The net is once more plunging its way to the bottom, the vessel glides through the rippling music, and you are absorbed in eager inspection of shell and weed. It is probable that this stooping and peering, accompanied by the motion of the vessel, will bring on the nausea and headache, if not worse, which hitherto you have escaped. I will not pretend that this is pleasant; but there is no help for it. None but the brave deserve the mollusc! The pain is transient, the delight persists. You may return home at the close of the day probably uncomfortable, and certainly hideous; but behind you Jack is bringing a bucketful of treasures; and to-morrow you will only know that you have these treasures.

The first thing you have to do on the morrow is to "identify" the animals—a long and interesting, though sometimes perplexing process, owing to the exasperating system adopted by naturalists of frequently selecting, as marks, characteristics by no means obvious. For example, when you read the sentence "shell flexible," among the curt indications by which an animal is to be identified, how are you to suspect that the animal in question has *no* shell visible at all, until you have dissected it, and found the thin calcareous plate underneath the back, covering the liver? That one sentence "shell flexible" prevented my identifying a *Pleurobranchus* for at least an hour.

Nor have I to this day been able to identify the species of a compound Ascidian (which I only know to be an Ascidian from embryological indications), probably known to naturalists, perhaps yet undescribed. It is of a bright orange colour. From a transparent gelatinous basis minute cylindrical tubes rise, each about the twentieth of an inch in height, standing in circular groups. The orifice of each tube has four delicate processes radiating inwards, like the spokes of a wheel, or like the processes in the siphon of a cockle. This orifice is alternately protruded and retracted, but does not open and shut like that of an Ascidian; and, moreover, the orifice is single. The heart, or pulsating sac, lies at the bottom of the visceral cavity. Imbedded in the clear gelatinous base are several branching vessels giving off pear-shaped processes. These vessels connect the visceral cavities of the whole colony, and the globules of food are seen oscillating to and from the cavities into the pear-shaped processes. I was completely puzzled what to consider this animal, until I saw a tadpole embryo escape from it, and swim away, followed by several others; and then I knew an Ascidian of some kind was before me.

A tadpole? Well, that is a figure of speech. The embryo of the Ascidian is more like a tadpole than anything else; and totally unlike its parent, not only in possessing a good long tail, but in being able to swim vigorously through the water in which the parent is immovable. In the interior of the round body which surmounts this tail, a mass of yellowish granules (the vitellus) is observed, which extends some way down the axis of the tail. The transparent membrane surrounding the granular mass enlarges. The mass develops three processes, which act as suckers, wherewith the animal finally fixes itself for life. The tail then becomes absorbed, as in the tadpole.* The viscera appear; the envelope increases, and finally becomes the general basis out of which, or in which, an immense number of Ascidians are developed by the process of " budding;" so that from this one tadpole embryo there arises a whole colony of animals, from which in turn solitary tadpoles will issue, each of which will produce its colony. Imagine a tadpole to be transformed into a mature frog, this frog to swell his skin

---

* Some writers describe this disappearance of the tail as a fission, the tail dropping off. I have not observed this. The enveloping membrane, as it enlarged, included the tail within it; and the absorption took place within the sac thus formed.

to an indefinite extent, and under that skin to produce, by budding, some hundreds of frogs, all living harmoniously together, each fed by all—further imagine this colony producing at last a few solitary tadpoles, and you will have some conception of the paradox presented by our compound Ascidians.

Nor is this paradox without parallels. The other day I noticed the surface of the water in my pan agitated, as if scores of hairs were at various points thrust upwards. Nothing else was visible with eye or lens. Suspecting, from a certain pulsating motion, that it was caused by young Medusæ, I dipped the zoophyte trough, and brought up a quantity of newly-hatched Medusæ in great activity. They had just issued from the Polype (*Laomedea geniculata*), and on removing some of the Polype branches to the microscope, the young Medusæ were plainly visible in the capsules, and were easily pressed out, whereupon they swam away like the others. (Plate IV., fig. 1, represents a Campanularian Polype with the young Medusæ in the capsule.) Familiar as this sight was to me, it had not lost its marvellousness. Here was a Polype, which the uninstructed eye could not distinguish from a seaweed, producing scores of Jelly-fish; and these Jelly-fish, if their days were spared, would in due time produce Polypes. Imagine a lily producing a butterfly, and the butterfly in turn producing a lily, and you would scarcely invent a marvel greater than this production of Medusæ was to its first discoverers. Nay, the marvel must go further still; the lily must first produce a whole bed of lilies like its own fair self, before giving birth to the butterfly; and this butterfly must separate itself into a crowd of butterflies before giving birth to the lily: when you have thus added marvel upon marvel, you will be ready to listen without scepticism to the phenomenon known as the "alternation of generations," since Steenstrup so baptised it. Others have given it other names: Owen calls it "Parthenogenesis;" Van Beneden, "Digenesis;" and Quatrefages, "Geneagenesis." But while differing about the name, and the explanation of the phenomena, there is no difference as to the phenomena themselves. I will ask the reader's attention to a succinct exposition of the various facts and theories connected with this interesting subject; premising that I have not only verified the capital observations on which the marvel rests, but have some new facts to bring forward which must materially modify the current conceptions.

Harvey's celebrated aphorism, *Omne vivum ex ovo* (every living being issues from an egg), was a premature generalisation, and has for some years past been known to be so. Many animals issue not from an egg, but directly from the substance of the parent's body, by a process analogous to that of the budding of plants. To include this process and the ordinary process under one expression, Auguste Comte suggested the following modification of the aphorism, *Omne vivum ex vivo* (every living being issues from a living being); and as the idea of spontaneous generation becomes every year less and less tenable, this aphorism acquires the force of a law. I allude to it at starting, because, inasmuch as the course of our inquiry will conduct us to the conclusion that Generation is *not* essentially a distinct process from that of Growth in general, the idea of an ovum as the necessary origin of every living thing needs to be modified. The first illustration we owe to Trembley, whose Memoirs on the *Hydra*, or Fresh-water Polype, are so admirable in accuracy and extent of observation, that, in spite of the labours of a century, nothing of what he stated has been set aside, and very little added, except what the microscope has revealed. He taught us that the Polype, which originally comes from an egg, produces a quantity of other Polypes, exactly similar to itself, by a process of " budding," after the manner of a plant. He taught us, moreover, that not only is this the normal mode of multiplication, but that if we lacerate the Polype, each lacerated fragment will become a new Polype, which in its turn may be cut into several pieces, every one of them developing into perfect Polypes. Several naturalists have repeated and confirmed his experiments. In repeating them myself I failed at first, but subsequently succeeded, and attribute the first failure to the presence of impurities in the water containing the fragments. Mr R. Q. Couch made the curious observation,* that if the body of the Hydra " be merely irritated with a needle, or a ray of

---

* *Reports of the Penzance Natural History Society*, 1850, p. 571. SCHLEIDEN says of the *Gesneria*, that a puncture in the leaf produces a bud in a few days. The two cases are precisely similar.

the sun, a young one will sprout from the injured parts." Here Harvey's dictum receives direct contradiction: the Polype which is produced from a wound in the body of the parent, being in every respect similar to the Polype which is produced from an egg.

It was in 1744 that Trembley made known to the world the astonishing reproductive powers of the Hydra.* The following year † Bonnet published his not less astonishing revelations on the reproduction of *Aphides*, or plant-lice. The *Aphis*, a winged insect familiar to most readers, deposits its eggs in the axils of the leaves of plants at the close of summer, and these eggs are hatched in the following spring; but the insect which issues from the egg is a wingless, sexless insect. It was known that this wingless insect brought forth its young alive. Bonnet proved that this took place when no male insect was present—in fact, he proved that the insect was a virgin mother, and astoundingly fertile. He isolated the young Aphis as soon as it was hatched, reared it in strict seclusion, and watched it daily, almost hourly, with the patient tenacity of a naturalist of genius. He has left on record his anxieties, his tremulous agitation lest its death should supervene to frustrate his labours; and his joy, after seeing the captive four times change its skin and reach its normal development, to observe that this absolute virginity did not in the least

---

* TREMBLEY: *Mémoires sur un genre de Polypes d'eau douce*, 4to, Leyden, 1744.

† BONNET: *Traité d'Insectologie*, 2 vols., 1745; vol. i., p. 26 *et seq.* A better edition is that printed at Amsterdam, 1780.

interfere with fertility. On the eleventh day the Aphis produced a young one alive; another succeeded, and another. Every four-and-twenty hours the brood was increased by three, four, and even ten arrivals. At the end of twenty-one days, ninety-five young ones were produced from this single Aphis.* Carrying further his observations, Bonnet found that the virgin offspring of this virgin parent also became parents!† We know that this reproduction often goes on till the eleventh generation: then this process ceases, the last generation is of *perfect* insects, with separate sexes, and these produce ova which next year become the productive virgins we have just been reading of.

"But why," we may ask, in the language of Professor

* "Enfin pour achever l'histoire de notre pucerone je n'ai plus qu'à dire qu'ayant été obligé de m'absenter pendant tout le 25 jusqu'au lendemain matin sur les cinq heures, j'eus le chagrin à mon retour de ne la pas trouver où je l'avois laissée, ni dans les environs où je la cherchai inutilement. Comme, depuis qu'elle avoit commencé d'accoucher je n'avois pas cru qu'il fût nécessaire de la tenir renfermée exactement, elle en avoit sans doute profité pour aller finir ses jours ailleurs. On juge aisément que je ne fus pas insensible à cette perte. J'avois vu naître cette pucerone, je l'avois suivie constamment pendant plus d'un mois, et je me faisois un plaisir de continuer à l'observer jusqu'à sa mort. Je me proposais de sçavoir au juste le nombre de pucerons dont elle auroit peut être encore accouché. Il y a apparence qu'il n'auroit pas été considérable, à en juger par l'extrême diminution de sa taille. Son ventre qui lorsqu'elle n'avoit fait encore que peu de petits, étoit arrondi et comme distendu, s'étoit aplati, et étoit devenu de forme triangulaire."—Vol. i. p. 49.

† "Si la petitesse de la pucerone de la troisième génération m'avoit surpris, j'eus lieu de l'être encore davantage de celle de sa fille. Elle ne sembloit pas avoir atteint la moitié de la grosseur qu'ont ordinairement les pucerones de cette espèce lorsqu'elles commencent à engendrer."—Vol. i. p. 78.

Owen, "should there be this strange combination of viviparous generation at one season, and of oviparous generation at another in the same insect? The viviparous or larviparous generation effects a multiplication of the plant-lice adequate to keep pace with the rapid growth and increase of the vegetable kingdom in the spring and summer. No sooner is the weather mild enough to effect the hatching of the ovum, which may have retained its vitality through the winter, than the larva, without having to wait for the acquisition of its mature and winged form, as in other insects, forthwith begins to produce a brood as hungry and insatiable and as fertile as itself. The rate of increase may be conceived by the following calculation. The aphis produces each year ten larviparous broods, and one which is oviparous, and each generation averages 100 individuals:—

Generation.	Produce.
1st,	1 Aphis.
2nd,	100, a hundred.
3rd,	10,000, ten thousand.
4th,	1,000,000, one million.
5th,	100,000,000, hundred millions.
6th,	10,000,000,000, ten billions.
7th,	1,000,000,000,000, one trillion.
8th,	100,000,000,000,000, hundred trillions.
9th,	10,000,000,000,000,000, ten quatrillions.
10th,	1,000,000,000,000,000,000 one quintillion.

"If the oviparous generation be added to this, you will have a thirty times greater result."*

* Owen: *Comparative Anatomy*, p. 414. [This calculation, however, has been utterly knocked to pieces by Huxley, in his Memoir on the Aphis, *Linnæan Trans.* xxii. 215.]

Recovering from the stupor into which we are thrown by facts like these, let us observe that here, as in the case of the Ascidians and Polypes formerly mentioned, an alternation of generations takes place; the parent producing a child *unlike* itself, and that child in its turn finally producing one like its grand-parent. " The winged and perfect Aphis produces a wingless hexapod larva; this wingless larva produces at last a winged and perfect insect." The reader may imagine how great was the sensation produced in the scientific world by these announcements, and how many theories were propounded in explanation; we must not pause here to consider them, but proceed with our history.

The last date was 1745. In 1819 a Germanised Frenchman, known to all lovers of romance as the author of *Peter Schlemil*, made a discovery in Natural History which was almost as incredible as his Shadowless Man. Whether this will endear the name of Chamisso still more to his admirers may be a question. Literary men will point with some satisfaction to the fact that a novelist was the discoverer of a form of reproduction unsuspected by the profoundest zoologists. They may also remember that the luminous doctrine of plant-morphology was the discovery of the greatest of our modern poets; and that the great Haller himself was a poet and *littérateur* before, in later life, he devoted himself with such splendid success to physiology.

In Chamisso's day, naturalists knew two distinct species of the curious mollusc named *Salpa*, an indescribable animal, transparent as crystal, and of irregular cylindrical aspect. This animal is also seen somewhat different in structure, but most obviously differing from the solitary species in being a long chain of animals. In spite of their differences, they are not two species, but two generations of the same species. The solitary Salpa produces the chain-salpa by " budding ; " and the chain-salpa by " alternation of generations " (the phrase is Chamisso's) produces the solitary Salpa by ova. Krohn, Huxley, Leuckart, and Vogt (alas! only one Englishman among four Germans), have since confirmed Chamisso's discovery, which, as Mr Huxley has pointed out, gives him the priority over Steenstrup, not only as to the mere phrase of " alternate generations," but as to the distinct conception of the idea implied in the phrase. Nine years afterwards, in 1828, Milne Edwards first announced a similar mode of reproduction among the Ascidians (such as I sketched it just now), without, however, connecting it with Chamisso's discovery. In 1835, the Norwegian pastor and indefatigable naturalist, Sars, opened that wonderful series of revelations which by himself, Loven, Lister, Dalyell, Steenstrup, Van Beneden, Allman, Forbes, and others, have established the alternation of generations in Polypes and Medusæ.

A not less surprising alternation has been discovered in the Entozoa ; but it would occupy too much space to narrate here, requiring much preliminary explanation before it could be intelligible to the general reader.* Let us continue our history.

* The student will find a complete history of these phenomena in

In 1842, the known facts were collected, and connected under one generalisation by the Danish botanist Steenstrup, who brought his own quota of important facts. In this work,* a flash of light suddenly revealed the connection in which many isolated paradoxes stood to each other: a theory was proposed, which, although really nothing but a metaphorical expression of the already known facts, was very widely accepted as a perfect solution of the difficulty.

In 1849, Professor Owen published his two lectures on *Parthenogenesis*, in which, re-stating the results of his investigations into the reproduction of Aphides (1843), he propounded a theory as a substitute for the metaphor of Steenstrup, and one which up to this time is the sole theory not open to the charge of being a merely verbal explanation. In the same year, Victor Carus published a small work† containing some new observations, and an ingenious classification.

In 1851 Leuckart published an essay‡ to prove that alternate generation was simply metamorphosis *plus*

Siebold, *Über die Band-und-Blasenwürmer*, 1854, or in the larger work of Küchenmeister, *Die in und an dem Körper des lebenden Menschen vorkomenenden Parasiten*, 1855; a translation of which, by Dr Lankester, has been published recently by the Sydenham Society. Van Beneden has since published his splendid monograph *Sur les Vers Intestinaux*, which gained the prize offered by the French Academy.

* *On the Alternation of Generations.* Translated for the Ray Society by Mr George Busk, 1846.

† *Zur nähern Kenntniss des Generationswechsels*, 1849.

‡ Siebold & Kölliker's *Zeitschrift*, iii. p. 170. He repeats the idea in his work on Comparative Anatomy, written in conjunction with Bergmann. [He has since published a valuable little work, *Zur Kenntniss des Generationswechsels*, 1858.]

asexual generation—an unhappy explanation, since the peculiarity of metamorphosis is that the larva becomes a perfect insect, whereas the Polype never *becomes* a Medusa, it only *produces* it; the wingless Aphis never *becomes* a perfect insect; moreover the phrase " *plus* asexual generation" conceals the real difficulty.

In 1853, Van Beneden, to whom we owe so many important contributions, published a work* in which he modestly contents himself with stating the phenomena, classing animals under two heads, *monogenetic* or sexual, and *digenetic, i. e.* reproducing themselves by sexual and asexual methods.

In 1855 M. Quatrefages published four articles in the *Revue des Deux Mondes* entitled *Les Métamorphoses*, in which he reviewed the state of the question, criticised the theories, and propounded one of his own.

In 1856 another brilliant flash of light came from Germany. Von Siebold published a work † containing some startling facts, and such as, in my opinion, will serve to dissipate all the clouds from the question. He offered no theory himself; and in the only remark which directly touches our subject, he desires to lay " particular stress upon the distinction between the alternation of generations and Parthenogenesis." In spite of this, I must think that the two are one, and that his facts convincingly prove them to be so. For the present,

---

* *La Génération alternante et la Digénèse*, 1853.
† *On true Parthenogenesis in Moths and Bees.* Translated by J. W. Dallas, 1857.

however, we will confine ourselves to the points established in his work bearing on our subject.

Having isolated female moths, he constantly watched them in little vessels closed with glass lids. In due time they laid eggs. There was nothing surprising in this; the virgin moth, as well as the female of every other insect—indeed of every other animal—lays eggs; but what was his astonishment, " when all the eggs of these females, of whose virgin state I was most positively convinced, gave birth to young caterpillars, which looked about with the greatest avidity in search of materials!" Imagine a brood of chickens hatched from the eggs of a virgin hen, and you will conceive Siebold's surprise. He subsequently found that bees, in like manner, produce hundreds of eggs, which, however, invariably become *male* bees; for it is only the fertilised bee-egg which will develop into a female—either worker, or queen.

Ungallant physiologists, resting on the evidence of some embryological phenomena, have declared the female to be only a *male in arrested development;* a very impertinent deduction, which was, however, flung back on them by a witty friend of mine, who, hearing that one of her own sex was fond of reading metaphysics, and was feared to be suffering from a softened brain, drew her own conclusion as to this masculine course of study, exclaiming, " *Man is but woman with a softened brain!*" She would have also retorted Von Siebold's facts about the bees, which point at a miserable inferiority on the part of the males. But I must not let her prematurely enjoy this triumph: if the imperfect bee is always a male, the imperfect moth (*Psyche*) is always a female; and to reconcile both parties, we have the silk-worm moth, whose virgin progeny is *both* male and female.

In conclusion, be it noted that Von Siebold's work establishes Parthenogenesis as a *normal* process in bees and moths, on grounds which, Funke justly says, do not permit the severest scepticism to raise a doubt worthy of notice.* He, moreover, points to the fact that among the Entomostraca there are species of which *only* the female is known; again, thousands of females of the gall-fly have been examined, but not a single male has yet been found.

Thus Parthenogenesis has been found to exist in Polypes, Molluscs, Annelids (by M. Quatrefages in the *Syllis*), and Insects. It has also been found in plants, The *Cœlebogyne Ilicifolia*, one of the *Euphorbiaciæ* to be seen in Kew Gardens, is a striking example. Only the female of this plant has reached England, yet it continues yearly to produce descendants, although no male has arrived.† "In fact," says Dr Lankester, "Parthenogenesis, in all its integrity, has now been observed in a large number of cases in the vegetable kingdom. The occurrence of seeds, independent of stamens, was first observed in a Euphorbiaceous plant in the gardens at Kew. It has subsequently been ob-

---

* FUNKE. *Lehrbuch der Physiologie*, 1857, p. 1326.

† RADLKOFER: *Ueber wahre Parthenogenesis bei Pflanzen*, in SIEBOLD u. KÖLLIKER, *Zeitschrift*, viii. 458.

served in a large number of plants, a list of which, with the observer of the phenomenon, we subjoin :—

CHARACEÆ—*Chara crinita.* A. Braun.
CANNABINEÆ—*Cannabis sativa.* Naudin.
CHENOPODIACEÆ—*Spinacea oleracea.* Le Cocq.
EUPHORBIACEÆ — *Cælebogyne Ilicifolia.* J. Smith; *Mercurialis*, species. Naudin.
ANARCARDIACEÆ — *Pistacia Narbonensis.* Tenore; *Pistaciæ*, species. Bocconi.
CUCURBITACEÆ—*Bryonia diocia.* Naudin.
DATISCEÆ—*Datisca cannabina.* Fresenius."*

To this list may be added the *Algæ*, in which Pringsheim has observed the same phenomenon.†

Such were the facts known at the time when I resumed my investigation of Polype-parthenogenesis. The labours of distinguished naturalists on the genesis of Polypes may be summed up in the following *schema:*

A. The Medusa parent *produces* ova;

B. These ova are *developed* into infusoria;

C. These infusoria are *developed* into Polypes;

D. These Polypes *produce, by budding,* the Medusæ, which in turn *produce* ova.

Thus D completes the cycle commenced at A. As *variations* from this route we have—

α. The Medusa produces Medusæ by budding;

β. The Polype produces Polypes by budding;

* LANKESTER: in *Microscopical Journal*, No. XX., July 1857.
† *Annales des Sciences Naturelles.* 1856.

γ. The Polype produces Polypes *by ova* directly, *i. e.* without going through the Medusoid generation.

Attention is called to this second series, because the facts therein registered have been too often lost sight of in the discussion of the theory. When, for example, so much stress is laid on the analogy between the development of a Polype into a Medusa, with that of a bud into a flower, it is apparently forgotten that, in spite of the resemblances, great differences are discoverable. No flower produces similar flowers by a process of budding, as the Medusa buds off young Medusæ from its substance: a rose does not split up into a dozen roses.

So little have the facts registered in the second table been kept in view, that the doctrine of Alternate Generations has been persistently denied on the ground that the Polypes are not generations at all—are not, properly speaking, "individuals" any more than leaf-buds are individuals. According to this argument, which has been set forth by Dr Carpenter,* only those can be truly called generations which issue from a generative act, *i.e.* the union of a germ-cell with a sperm-cell; an arbitrary assumption disproved by a multitude of facts. He maintains the analogy of the Polype and the leaf-bud to be complete, and considers the multiplication of Polypes, and of Medusæ from Polypes, to be always a process of budding; this gives his argument a superficial plausibility, which is, however, totally destroyed by the fact that the Polype *also* produces Polypes by the union of ova and spermatozoa.

* *Principles of Comparative Physiology*, 1854.

I shall have to recur to this point hereafter; meanwhile I may add that, in the course of a long investigation into the development of the Campanularian and Plumularian Polypes — especially *P. myriophyllum* (Plate IV., fig 2), from deep water off Jersey—I found that not only does the Polype produce Polypes by means of ova, but *also produces Medusæ in the same way;* so that instead of the production of Medusæ being only one of simple budding, it resembles the production of Polypes in being *sometimes* a process of budding, and *sometimes* a process of oviparity. I have followed this development through all its stages; and as what I have seen may be seen by any one who chooses to devote the requisite patience, I shall merely clear away certain theoretical obstructions which may screen the real facts.

In Dr Carpenter's summary of the views held by naturalists, we read that the ovarian capsules (the large vesicles which rise from the stem of the polypidom) are improperly designated ovarian, because " they have been shown by Professor E. Forbes to be in reality metamorphosed branches." The force of this objection escapes me. Wolff and Goethe have shown the stamens and pistils to be metamorphosed leaves, but no one denies them, on that account, to be reproductive organs. The capsule in question is *not* a branch, but a capsule, and the proof of its being an ovarian capsule is the fact that in it ova are developed. This, indeed, Dr Carpenter denies, for he continues,* " These Medusa

---

* *Principles of Comparative Physiology*, p. 552.

buds spring not from ova, but from a detached portion of the medullary substance;" and in a note he adds, "Although they are described by Van Beneden as developed from ova, yet it is clear from his own account that such is not the case; and that what he called the vitellus is continuous with the medullary substance of the stem and branches of the zoophyte." Not having seen Van Beneden's *Mémoire*, I am unable to say whether that admirable naturalist has imperfectly described what he has seen, or Dr Carpenter imperfectly comprehended what he has read; but I have no hesitation in asserting that direct study of the phenomena will disclose the fact of the Medusa being, at any rate, *sometimes* developed from ova, *although* the vitellus is "continuous with the medullary substance of the stem." The ova are there, unmistakable by any eye familiar with the ova of zoophytes; and by cutting off the tips of the capsules we can gently press these ova out, revealing the germinal vesicle in each, and the vitelline mass surrounding it. Not only are ova there, but in some instances spermatozoa may be observed in great activity, and this at a time when the circulation, or more properly *oscillation*, of medullary granules from the stem into the interior of the capsule is perfectly visible. Sometimes, instead of these, we find simply a mass of granules and nucleated cells; at other times, ova in various stages of segmentation, the germinal vesicle having disappeared, and a vitelline membrane being formed; at others, we find embryos nearly ready to escape.

The facts which I have observed are so opposed to the current theories on this subject, that I have no expectation of their gaining acceptance, until they have been confirmed by others. But I am content to await that confirmation, assured that it must come sooner or later. Scepticism, be it never so authoritative, cannot alter the facts; and as I am quite sure of what I have seen, and seen many times, I am sure it will be seen by others. Unfortunately the *Myriophyllum* is only to be had in deep water, and appears not to be common, although it is abundant off Jersey; and there is also a *possibility* that the phenomenon may not be observable in this species, dredged from other coasts. I say *possibility*, because I am informed by Professor Kölliker that the *same* species of Polype (*Eudendrium*) found off our coasts and the coast of Naples, differs in this remarkable character: the one produces eggs, and not Medusæ; the other Medusæ, and not eggs. The same great anatomist assured me that he also had seen something like what I have seen; he had seen the same species of *Campanularia* producing both eggs and Medusæ, "but no one believed me," he added. The plea for scepticism being, that Professor Kölliker must have mistaken two different species for the same species — a hazardous assertion, considering *who* is the observer. But this objection cannot apply to my observation, for it was on the *same Polypidom* that I found some of the capsules filled with eggs, and some with Medusæ. Even more startling is the fact now to be mentioned. In one and the same species (*Myriophyllum*), dredged at the same time and from the same place, I found the *capsules containing eggs and also Medusæ*; and others—but not on the same Polypidom—*containing eggs and Polypes*, *i.e.* the ciliated gemmules which we know to be the infusorial stage of the Polype.

I have seen this so often that the whole history of evolution thus presents itself: Taking the medullary substance of the Polype as the analogue of the cellular basis of the plant, we may trace a somewhat similar course of evolution in each: the cellular basis becomes differentiated into leaves, stamens, pistils, germ-cells, and sperm-cells; the medullary substance becomes differentiated into nucleated cells—these cells into germ-cells and sperm-cells, or into *germ-cells alone*, from which are developed,

1st, Under one set of conditions, probably of temperature and food, Polypes;

2d, Under another set of conditions, Medusæ. Just as a leaf-bud is developed under one set of conditions, and a flower under another set; or as only germ-cells are developed in one plant, sperm-cells in another, or both on the same plant.

Of great importance as regards the theory of Parthenogenesis are two of the facts just indicated—namely, that the Polype produces ova which become Medusæ; and that these ova may indifferently become either Polypes *or* Medusæ. The latter fact ceases to be so marvellous, when we consider that Agassiz has demonstrated the identity in structure of Polype and Medusa.

Of still greater importance as regards the theory of Parthenogenesis is the conclusion that from *germ-cells only*, without any influence from sperm-cells, Polypes and Medusæ may be developed. Do you ask for evidence on which to base this conclusion? The evidence is various. We have already noted the indubitable fact that the unfertilised eggs of Entomostraca, gall-flies, bees, moths, and silk-worms, *do* become developed animals; these must spring from germ-cells only. Next we have the recent experiments in France and Germany, which place beyond a doubt that diœcious plants become fertile when their pollen is entirely removed; and we know that female plants become fertile when no pollen-bearing plant is in the kingdom; so that Schleiden, in the edition of his work, *The Plant*, just issued, says, "We must now confess complete ignorance as to the real function of the pollen."[*] To these may be added the fact that, according to my observations, spermatozoa are rarely met with in the Polypes, whereas ova are extremely abundant.

All this evidence, however, is as nothing beside that which is furnished in the very remarkable researches on the reproduction of Aphides made by Professor

---

[*] SCHLEIDEN, *Die Pflanze*, 1858, p. 72. He draws the piquant conclusion that it is precisely in those plants on which Linnæus founded his sexual system that we have now *no* evidence of sexuality, whereas in the cryptogamic plants sexual distinctions have been accurately ascertained. It seems to me, however, that this paradox is more piquant than true, and that the analogous phenomena of Parthenogenesis in animals explain the difficulty, without forcing us to deny the sexuality of the Phanerogamia.

Huxley, an account of which he read to the Linnæan Society, Nov. 5, 1857. That paper not being yet printed, I can only refer to its conclusions in as far as they bear on the present topic, and these are—First, that the virgin viviparous Aphis, which is said to produce broods of young by "internal gemmation," does really produce them *from unfertilised ova:* every stage of the development of these ova, and that of the embryos having been carefully made out. Secondly, that the female oviparous Aphis produces her young from ova fertilised by spermatozoa; but the early stages in the development of these ova are *precisely similar* to those of the viviparous Aphis, the differences presenting themselves at a later period, when in the one case the formation of the "mulberry mass" takes place spontaneously, in the other case it is preceded by the union of the spermatozoa with the ova. As the product of both processes is identical—as the Aphis which is seen to issue from an unfertilised ovum is in all respects similar to that which issues from a fertilised ovum—the conclusion is irresistible, that Generation is possible from germ-cells only; and the facts I have brought forward respecting the Polypes are thus seen to stand in no absolute isolation.

Resuming in one schema the results of my investigations into Polypes, with those of my predecessors, we find,—

A. The Medusa parent produces ova;

B. These ova are developed through an infusorial stage into Polypes;

C. These Polypes, in turn, produce ova;

D. (1) These ova are developed into Medusæ, thus completing the cycle opened at A.

D. (2) These ova are developed into Polypes, thus completing the cycle opened at C.

The budding process, which both Medusa and Polype manifest, may be eliminated from the scheme of "Alternation." We shall hereafter see that it is essentially the same as the other processes of generation.

Such, in brief, is the history, such are the facts, of Parthenogenesis. Let us now glance at the theories which attempt to explain them. Steenstrup—whose merits are very considerable, and who first propounded a general theory, named by him the "Alternation of generations"—encumbered the question, instead of clearing it, when he called the Polype the "wet-nurse" of the Medusa, denying its claim to be considered as a "parent." To say that the Polype is not properly a "parent," but has only the germs of the Medusa confided to it, is, as Professor Owen justly remarked, to make a metaphor supply the place of an explanation. In reply to this objection Steenstrup boldly declares his theory is *la combinaison intime des faits.* Professor Owen convincingly shows that the theory is purely verbal; and I would hold that it is in direct antagonism with the fact that the Polype sometimes produces eggs without the mediation of a Medusa; and if a Polype, issuing from an egg, and also producing an egg from which another Polype will issue, be not regarded as a "parent," it will be difficult to specify in what parentage truly consists. Steenstrup's theory is almost identical, except in language, with that of the old writer alluded to by Quatrefages, who accounted for Bonnet's facts by a "transmitted fecundation:" " D'après lui, les pucerons produisent toujours des œufs aussi bien que les autres insectes, mais chez eux la fécondation, au lieu d'agir sur une génération seulement, étend son influence à plusieurs générations successives. Elle devient par conséquent inutile jusqu'au moment où la somme d'action transmise de mère à fille est totalement épuisée."

At a first glance this may be mistaken for an anticipation of Owen's theory; but a more rigorous inspection discovers that Owen's theory differs from it by the all-important character of definiteness. Instead of throwing over the question the obscure generality of a phrase, it points directly to a specific fact, or condition, such as, if accepted, would indicate the terminal stage of inquiry, beyond which no intellect could hope to penetrate. It starts from the germ-cell, from which the organism arises, and, following the course of this germ-cell, it holds the Ariadne thread, which, through all the mazes of the labyrinth, conducts the mind to clear issues. Let us, in as brief a space as possible, develop this theory.

Every organism, plant or animal, originates in a cell. This cell spontaneously divides into two, these two into four, these four into eight, and so on, till, instead of a solitary nucleated cell, a mass is present, known as the "germ mass." (Plate VI., figs. 4, 5, 6, 7, 8.) In the *Conferva*, instead of a mass, a thread of cells has arisen, forming the filament which constitutes the whole plant. In the animal, the cells are not placed end to end, thread-like, but side to side, and form what is called the "mulberry mass" (fig. 8); and a further distinction is to be noted, namely, that each animal cell, as it formed, carries with it a portion of the yolk. From the "germ mass" the animal is evolved. Each cell of this mass is the offspring of the primary germ-cell, reproducing its powers and capacities. Since the animal is formed out of this mass, and by means of it (figs. 9, 10), we are forced to the conclusion that the cells have become transformed into tissues. But "not all the progeny of the primary germ-cell are required for the formation of the body in all animals : certain of the derivative germ-cells may remain unchanged, and become included in that body which has been composed of their metamorphosed and diversely combined or confluent brethren : so included, any derivative germ-cell, or the nucleus of such, may commence and repeat the same processes of growth by imbibition, and of propagation by spontaneous fission, as those to which itself owed its origin."*

It is this, according to Owen, which constitutes Parthenogenesis. Some of the cells, instead of being transformed into tissues, remain, unchanged as cells, included in the body, where they repeat the original process of subdivision, and produce offspring as they

* OWEN: *Parthenogenesis*, p. 5.

themselves were produced. In proportion, therefore, to the complexity of the animal (that is, in proportion to the amount of cells transformed into tissues), will be its inability to reproduce itself by Parthenogenesis. In proportion to the amount of unchanged cells will be this power of reproduction. The marvels of the *Hydra*, as recounted by Trembley, are thus explicable ; for the Hydra retains its germ-cells unchanged everywhere, except in the tentacles and the integument, and *these are incapable of reproduction.* " The reproduction of parts of higher animals has also been found to depend on pre-existing cells retained as such. Mr H. D. S. Goodsir has shown that, in the lobster, so noted for the power of reproducing its claws, the regenerative faculty does not reside at any part of the claw indifferently, but in a special locality at the basal end of the first joint. This joint is almost filled by a mass of nucleated cells surrounded by a fibrous and muscular band."

But here the reader may ask how the cycle of generation is ever completed? Why does not the Polype continue budding off fresh Polypes for ever ; why does not the Aphis-larva continue producing broods of larvae ; why does not the plant persist in sending forth leaves and buds ; why do we always see a sudden change—a leap, as it were, into higher life—completing the cycle by the Polype producing a Medusa, the larval Aphis producing an Aphis, the plant producing a flower? To this question Owen has prepared an answer. The original cell, in its frequent subdivision, gradually loses by dilution a portion of its plastic force. If on starting it had a force of 100, after fifty subdivisions it will have no more than 2. It is this necessary dilution of power in repeated reproductions which prevents Parthenogenesis from being indefinitely prolonged.

Such is the theory, in every way remarkable, proposed by our great anatomist; and before proceeding to examine its stability, I will adduce the strongest illustration in its favour I have yet found. The theory assumes that some of the original germ-cells are retained untransformed in the body of the Hydra and Aphis, which cells, in virtue of their original tendency, subdivide and develop into new animals. We have formerly seen that the germ-mass of the *Eolis*, *Doris*, and *Aplysia*, normally develops itself into one, two, three, and even eight distinct animals. As this takes place contemporaneously, and in the same chorion—as one egg actually divides into several embryos, by a simple process of subdivision in the germ-mass—I do not see how Owen's position can be denied, that here at least the offspring of the original cell is actually included in each distinct mass, and that it is the origin of each embryo. Whether the cells are *unchanged* or not, may be a question; it is certain that they are included: and as there can be little difference in the process, whether the progeny of one cell be developed simultaneously as in the Doris, or successively as in the Aphis, the fundamental position seems secured. I say *seems*, because I do not really think it is, nor do I find myself able to accept Owen's explanation.

Quatrefages and Siebold object to the name of Parthenogenesis as embodying an error. The larval Aphis, says the former, cannot properly be styled a virgin, because it is an incomplete organism, and "à l'idée de virginité se rattache invinciblement celle de la possibilité de cessation de cet état." He objects, therefore, to the name, because, he says, Owen's conception rests on the remarkable exception of the Aphis-larva, in which reproductive organs, incomplete, but still perfectly recognisable, have been discovered. The objection, which was never very forcible, is completely silenced by Von Siebold's discovery of perfect insects, male and female, in the virgin-progeny of bee and moth. As to Von Siebold's objection to the name, that by it Owen "confounds Parthenogenesis with alternation of generations," it is met not only by the explanation Owen gives in a note to the translation of Von Siebold's work (p. 11), but is further met by what will probably be seen, in the following discussion, to be the true state of the case; namely, that the generation of bees and moths is essentially the same as that of Ascidians, Aphides, and Polypes; and instead of confounding two distinct things in one phrase, Owen has reconciled two seeming differences.

Retaining, therefore, the name Owen has given to the phenomenon, let us examine his theory. Quatrefages, among objections of little weight, urges one of more value when he says that the process of segmentation in the yolk is now known to be different from that stated by Owen, being the spontaneous act of the ovum, whether the ovum be fertilised or not ; and farther, that the " yolk cells " are not cells at all. On this latter point it may be observed that embryologists are still divided,* the dispute turning on the correct definition of a cell—much as if men disputed whether a book "in sheets" ought properly to be called a "book." As regards Owen's theory, a slight modification in its terms would meet the objection.

Not so the objection which must, I think, be raised against the vital point in the theory—the assumption of a definite prolific force contained in the primary germ-cell, a force which becomes diluted by subdivision of the cell, and can be renewed only through another act of fertilisation. This is the heel of Achilles ; if vulnerable here, our theorist may be pricked by any vulgar javelin. Let us try. "The physiologist," says Owen, " congratulates himself with justice when he has been able to pass from cause to cause, until he arrives at the union of the spermatozoon with the germinal vesicle as the essential condition of development—a cause ready to operate when favourable circumstances concur, and without which cause those circumstances would have no effect. What I have endeavoured to do has been, to point out the conditions which bring about the presence of the same essential cause in the cases of the development of an embryo from a parent that has not itself been impregnated. The cause is the same in kind, though not in degree ; and every successive gene-

---

* See the latest work on the subject : FUNKE's *Lehrbuch der Physiologie*, p. 1366 *et seq.*

ration, or series of spontaneous fissions of the primary impregnated germ-cell, must weaken the spermatic force transmitted to such successive generations of cells."

Quatrefages justly calls this a seductive theory; but adds, that not even the imposing authority of Owen's name has gained acceptance for it. The first objection I should raise is, that the assumption of the "prolific force" belongs to *meta*-physiology. The second objection is, that it obliges us to embrace the paradox of the greatest effect arising from the most diminished force, since, according to it, the seed, in its primal vigour, only produces buds—in its exhaustion, flowers; the egg, in its primal vigour, only produces Polypes or Larvæ—in its exhaustion, Medusæ or perfect Aphides. Or must we regard the Flower, Medusa, and perfect Aphis as inferior and arrested forms, of which Leaf, Polype, and Larva are the matured beings? The celebrated Wolff maintained that the Flower was an imperfect organism—flowers and fructification, according to him, being the consequences of arrest of development;* and much may be said for this hypothesis, although we must finally reject it, when we know that there are plants which flower *before* they put forth leaves, and that the larval Aphis is confessedly an imperfect insect.

A third, and far more fatal, objection is, that under suitable conditions the plant will *continue* putting forth buds, the Polype putting forth Polypes, and the larval

* WOLFF: *Theorie von der Generation.* 1764. § 80, *et seq.*

Aphis putting forth larvæ, to an indefinite extent. The "prolific force," instead of diminishing, by repeated subdivisions of the cells, retains its primitive fertility. Kyber kept a plant, with larval Aphides, in a room the temperature of which was *constant*, and saw these larvæ produce broods for four years without interruption! Whereas, had the temperature varied, these larvæ would have manifested changes similar to those observed in ordinary circumstances, when the lowering of the temperature in autumn stops the production of larvæ, and induces that of perfect insects. We may also refer to the observation of Sir J. G. Dalyell, who kept a *strobila* for several years continually budding.

A fourth and last objection is, that the Polypidom, which produces both Polypes and Medusæ by gemmation, *also* produces eggs which become Polypes, as every one knows, and eggs which become Medusæ, as I have found; yet, *after* one of these egg-capsules has been developed on the Polypidom, the budding process continues as before. This would imply that the original prolific force, when nearly exhausted, produced eggs, and then, suddenly recovering its vigour, continued the production of buds. Now, an oscillating force of this kind cannot be accepted.

Although I think Owen's theory must be abandoned, it seems to me the best which has been offered—indeed, the only one which goes deeper than a phrase, and rests on definite conditions. The very definiteness of these conditions enables them to be closely tested and confronted with fact. The pregnant ideas contained in his work have been of essential service in the formation of those conclusions which force me to regard Parthenogenesis as not presenting any *peculiar* mystery. I shall endeavour to show that it is no *deviation* from the ordinary processes of Reproduction, except in formal and quite accessory details. Do not, however, suppose that, in denying the relative marvellousness of a phenomenon which has excited so much astonishment, any attempt is made to lessen the original marvel. When the rise of a feather in the air is explained by the same law of gravitation which explains the fall of the quill, no mystery is dissipated by this reduction of two seemingly contradictory facts to one law. In like manner, the eternal mystery of Reproduction remains the same dark Dynamis, baffling all comprehension, although by its laws we may also explain this novel phenomenon of Parthenogenesis.

Hitherto physiologists have admitted three forms of Reproduction. 1. The *fissiparous;* e. g. when a cell spontaneously divides into two cells. 2. The *gemmiparous;* e. g. when a plant puts forth buds, or a polype sends forth polypes from its stem. 3. The *oviparous;* e. g. when the plant and animal produce seeds and eggs. Fission, Gemmation, and Generation, are the three names designating these processes. The two first are universally admitted to be identical processes; but almost all writers regard Gemmation and Generation as two *essentially distinct* processes. Owen denies that there is any essential distinction. The Hydra, as he remarks, produces Hydræ both by Gemmation and

Generation. "The young Hydra from the bud is identical in organic structure and character with that which comes from the ovum; and when the effects of organic development are the same, their efficient causes cannot be 'altogether distinct;' only the non-essential accessories of the process may be the subject of variation." And Dr Alexander Harvey reminds us on this point, that "things that are equal to the same, are equal to one another;" so that if the product of the bud and the product of the seed be in all respects identical, there must necessarily be an identity between the seed and bud.* The potato-seed and the potato-bud both germinate apart from the parent plant, and both give rise to organisms in all respects identical. "What, let me ask, is included in the statement that the bud can evolve a perfect and complete plant—that it can evolve the flower and the seed? This: that it must *contain within itself* the *two kinds of cell* regarded as essential to the constitution of the seed,—as forming the essential characteristics of the seed, namely the 'sperm-cell,' and the 'germ-cell.'" † I do not accept the current idea respecting the germ-cell and sperm-cell as essentially *necessary* to the seed, and Dr Harvey himself seems to have relinquished the idea in his more recent work; but in both works he has established the essential identity of bud and seed, and consequently of Gemmation and Generation. Wolff long ago taught that the bud was

---

* See his two interesting works, *Trees and their Nature*, 1856; and *The Identity between the Bud and the Seed*, 1857.

† HARVEY: *Trees and their Nature*, p. 185.

identical with the seed;* but no one, I believe, has carried this doctrine to its legitimate conclusion, namely, that Generation is only a form of Growth; because every one has assumed that the union of two dissimilar cells is the necessary commencement of every generation. Even Owen, who maintains that Gemmation is closely allied to Generation, does so because he maintains that the original unchanged cells, which resulted from fecundation, form the starting-point of the bud; and that thus the bud and seed are identical, because both really originate in identical cells—both really issue from an original act of fecundation; whereas these pages contain abundant evidence that fecundation is by no means necessary to Generation, except in the higher animals; plants, polypes, insects, and crustaceans, being generated *without* fecundation.

It is worthy of remark that, although the Hydra propagates by eggs and by buds, it only produces two or three eggs during the autumn, whereas it buds all the year round. We may consider its oviparity, therefore, as an exceptional process. I believe it is one solely determined by external conditions, and that if the Hydra were kept in an unvarying temperature, it would never produce eggs at all, but continue budding to the end of the chapter.

All the endeavours to prove that Parthenogenesis is in every case the result of mere Gemmation are powerless against Owen, who denies the essential difference between Gemmation and Generation, and serve to sup-

* WOLFF: *Theorie von der Generation*, p. 47.

port his view when they are coupled with Von Siebold's discoveries. The Hydra sending forth a second Hydra from its own substance directly, may be said to "bud" like a plant. The Aphis producing broods of Aphides *internally*, instead of externally, which broods are unattached to their parent, may likewise be said to exhibit "internal Gemmation,"—although this budding is the result of ova. Von Siebold's virgin moths present us with eggs instead of young—eggs in every way identical with those produced by Generation: yet, if this be so, how shall we name the process? We must name it *internal oviparous Gemmation;* and what distinction there is between oviparous Gemmation and oviparous Generation, it will be difficult to say. In both cases, eggs are produced directly from the substance of the parent; these eggs, in *both* cases, develop into animals indistinguishable in structure or function, and capable of reproducing their species by *either* mode. From attending to formal and accessory differences, and not keeping the attention fixed on essential processes, physiologists have imagined a distinction to exist between Gemmation and Generation, which will not withstand scrutiny. Thus M. Quatrefages says : " In the animal, as in the plant, reproduction by budding is effected on the spot (*en entier sur place*), at the expense of the parent's substance. In the two kingdoms, reproduction by seeds and eggs demands the concourse of two elements prepared by special organs. It is immaterial whether these organs are both united in the same individual, or borne by distinct individuals ; there is always a father and a mother, a stamen and a pistil, an element which fertilises, and an element which is fertilised." The contrast is only formal. Out of the substance of the parent both bud and seed are evolved; whether the product shall be a mass of cells which at once develop into an *organism* by repeated subdivision, or into *eggs* by repeated subdivision, will depend on specific conditions; but the essential process is the same in each. The egg itself is a product, as much as the embryo; it is not a starting-point, but a station on the grand junction-line of development. No one will venture to assert that the process of Nutrition is other than identical, whether the product evolved from the blood-plasma be a nerve-cell, a muscle-cell, or a gland-cell: different as these products are, they all issue from embryonal cells indistinguishable from each other; and the law of Nutrition by which they increase is the same law in all.

The identity of the process in Reproduction is clearly seen in the following results of Mr R. Q. Couch's observations on the Sertularian Polypes: "At certain seasons of the year they produce cells (capsules) much larger than those of a more permanent character. These, at first, are composed of the granular pulp of the stem; afterwards the pulp becomes furrowed, and finally formed into cells. After a short period they separate from the parent, and undergo the process of development. If these cells attain a certain size, they are developed into eggs; if they are stunted by cold, they are formed into Polypes; while if, from unfavourable causes, they are still smaller, they grow into branches; and thus we see that, according to circumstances, different organs are capable of being eliminated from the same structure."* In conclusion, let us remember that the egg itself is an out-growth, not a starting-point; as all know who have made themselves acquainted with the results of embryological research, in which the phases of the genesis of the egg are minutely recorded; this genesis being the same essential process observed in *all other* forms of Growth.

Let us now examine the old position, which declares that the union of two different elements, a germ-cell and a sperm-cell, is the act of Generation—an act *sui generis*, and altogether distinct from the act of cell-multiplication, or Growth, which is regarded simply "as a modification of the nutritive function." The act of union, hitherto regarded as the fundamental act of all Reproduction, is only, I believe, a subsidiary, derivative process, and not by any means the "ultimate fact" at which our researches must pause; a conclusion to which Goethe pointed when he showed that Growth and Reproduction in plants are but different aspects of the same law.

Let us consider the known facts of Reproduction in their ascending order of complexity. What is the simplest process known? It is that of a cell spontaneously multiplying itself by subdivision. In the albuminous and starchy fluid, named *protoplasma*, a single cell appears. It assimilates more and more of

* *Penzance Nat. Hist. Society Report for* 1850, p. 374.

the fluid. It then divides into two cells perfectly similar. These two cells divide into four, eight, sixteen, and so the multiplication continues, till there is a filament of cells, each independent and capable of separate existence, but each attached to the other by its cell-wall. In a similar way leaves, instead of filaments, are formed. Many of the lower plants are nothing but aggregations of such cells; and in many this simple mode of Reproduction is the only mode yet discovered. By this process of subdivision a single cell of the *Protococcus nivalis* (or red snow) will redden vast tracts of snow in a few hours; and the *Bovista giganteum* is estimated to produce in one hour no less than four thousand millions of cells. Ehrenberg computes the increase of the infusorial *Paramecium* at two hundred and sixty-eight millions in a month. In this, the simplest form of Reproduction, the identity of the process with that of Growth is indisputable and undisputed.

The whole organism consists of a simple cell, or string of such cells: we must therefore either deny that the union of two dissimilar cells is the essential process of Generation, or we must point-blank deny that these cellular organisms are generated at all. If, shrinking from this latter alternative, we acknowledge that Generation must take place in these organisms, how shall we establish a line of demarcation between the reproduction of independent cells, and that of cells united together in a filament? In other words, how shall we demarcate Reproduction from Growth? When the cells are *attached*, a filament is formed, and the plant is said to grow; when the cells are *detached*, a new plant is said to be generated: but whatever differences there may be between twenty cells forming a filament, and twenty cells existing separately, each capable of growing into a filament, the *origin* of both is one and the same, and the process of Growth is identical with the process of Generation.

In a former passage I suggested that it was probably owing to differences of temperature, or food, that Reproduction by Gemmation and by Generation took place. This, which was hypothetical as regards the Polype, can be demonstrated in the Yeast plant. There are two kinds of yeast, or rather two forms of the same plant. The one is called *surface* yeast, the other *sediment* yeast. The former requires a temperature of 70° to 80° Fahrenheit; the latter 32° to 45°. Under the microscope we can watch the process of Reproduction in each. The surface yeast grows by budding only: from the cell-wall a little hernia is formed, which grows and grows, until in lieu of one cell there are two; these two set up the same budding process, and a whole filament of cells is the result. The sediment yeast does *not* bud; its isolated cells burst, and liberate a quantity of nuclei (spores?) which develop into perfect cells. If, however, the temperature be raised from 45° to 70°, this process is arrested; no more spores are formed, but the plant begins budding like surface yeast. Here, by a simple change in one of the conditions only (that of temperature), we convert indisputable Gemmation into indisputable Generation—unless we obstinately refuse to consider any reproductive process as a true generative act that is not preceded by the union of sperm-cell and germ-cell—a refusal which would lead to the denial of Generation altogether in vast regions of the vegetable and animal kingdom.

Having established this point, let us ascend a step, and we reach the second form of Reproduction, which is the *union of two similar cells*. This is named by botanists the act of "conjugation." In a simple filament, consisting of cells produced by fission, any two cells may unite; their contents coalesce to form a new starting-point, from which the multiplication of cells may proceed. Instead of two cells in the same filament, two cells of contiguous filaments may coalesce, but in each case it is the union of two similar cells. This is the first dim indication we obtain of that union of different sexes which in higher organisms becomes the normal process.

From the fission of one cell into two similar cells, and the conjugation of two similar cells, we now pass to the third and final mode of Reproduction, namely, the *union of two dissimilar* cells. To this union the special name of Generation has been applied; but the difference of name must not be allowed to mask the identity of the process. It is a fact, that for the production of the more complex organisms, union of germ-cells and sperm-cells is indispensable. Speculative physiologists have likened this union of germ-cell with sperm-cell to the union of an acid with its base. But the deeper our researches penetrate, the more erroneous does such a comparison appear. I cannot pause here to trace the genesis of ovum and spermatozoon, but must content myself with the assertion, which the reader can verify by consulting any embryological authority, that in their origin, and in the earlier phases of their development, these two cells are identical. It is only in their subsequent history that they differ.* If one convincing argument be needed to crown all these indications, we may find it in the now indubitable fact, that animals which normally are developed from fertilised eggs, are *also* normally developed from eggs unfertilised. It is clear, then, that if the egg, previous to fertilisation, has within it the elements and conditions *which will produce the same animal as would have issued from the fertilised egg*, the influence of the sperm-cell on the germ-cell, whatever it may be, cannot be of that elementary indispensable nature which is implied in the comparison of an acid uniting with a base to form a salt. No alkali spontaneously develops into a salt; without the acid the alkali is powerless to assume any of the saline forms. But the germ-cell does develop an embryo without the aid of a sperm-cell; and this, too, in certain animals which at other times generate sperm-cells. Indispensable the influence of the sperm-cell is, in the more complex organisms (although the insect is a very complex organism); but

* That is the reason why plants can be developed into male or female according to the will of the experimenter.

we observe one intensely significant fact, namely, that the *germ-cell spontaneously passes through the same early phases of its development, whether it be fertilised or not.* The germ-cell of a bird or mammal cannot *continue* its development, as the germ-cells of Polypes, Entomostraca, Bees, and Moths continue theirs; but neither is there any *fixed limit* to its arrest. Some ova fall short at one stage, others at others, but at no stage of their history can we say, Here the aid of fertilisation begins. *Every ovum, therefore, of the highest animal as of the lowest, has within it the power of development unaided by the spermatozoon*; this development falls very short indeed of an embryo, in the highest animals, but it travels some miles on the road towards that goal; and when, as in insects, the goal is not very distant, it may be reached. We may liken the spermatozoa to the extra pair of horses put to the carriage to enable it to reach a certain distance over mountainous ground. Two horses have dragged the carriage to the foot of the hill, and have brought it by precisely the same route as the four horses would have taken; but here, at the foot of the hill, the extra horses are indispensable. In granting the indispensable nature of the aid of such extra horses, no one would think of saying that it proved the necessity of four horses to carriage travelling.

The various modes of Reproduction we have seen to be identical, since not only are Fission and Gemmation admitted to be identical, but we have further seen that between Gemmation and Generation no real vital distinction exists. By a real and vital distinction, I mean one which implies an essential and indispensable difference in the two processes, and in the two results. But if the Hydra produced from a bud is in every respect the same as the Hydra produced from an ovum, and is capable in its turn of producing buds and ova, we can hardly suppose this identity of result to arise from processes essentially dissimilar. If the Bee, or Moth, produced from an unfertilised egg is precisely similar to the Bee or Moth produced from the fertilised egg, and is capable of producing offspring in the same way, we can hardly suppose this identity of result to arise from processes essentially dissimilar. We may distinguish the process of the *union* of two cells from that of the simple division of one cell, and call this union by the name of Generation; and there will be obvious convenience in having such a name; but if the *result* of Generation is the production of an animal perfect in all its parts, and capable of propagating its species, it is quite clear that the *union of two dissimilar cells* is *not* the essential and fundamental process necessary for such a result, since the result is frequently attained without it. When we consider Generation in the higher animals, we seem justified in establishing the *union* of germ-cell and sperm-cell as the distinctive and indispensable condition; but when we consider Generation in the abstract, and observe its phenomena in the simpler animals, we are forced to admit that this condition is no longer distinctive and indispensable, but that the union of the two cells is a secondary and derivative process, not the fundamental process of Generation.

In the simpler animals we have seen that no distinction whatever exists between Reproduction and Growth; and if in the more complex animals Reproduction is not carried on by this process of cell-division or cell-formation, the union of two dissimilar cells being indispensable, so likewise in those animals Growth is carried on by a more complex process. A vertebrate animal does not reproduce itself by spontaneous fission, like a conferva or an animalcule; but neither does a nerve grow by spontaneous fission.

Unless I am greatly deceived, the foregoing survey of the various forms of Reproduction has shown that there can be no *essential* distinction between Growth and Gemmiparous Reproduction. This granted, it likewise follows that as Gemmation and Generation are identical, there can be no *essential* distinction between Growth and Generation, but only formal accessory differences. Whether cells are aggregated together in filaments, or are set free as individuals, whether the cells develop into tissues, or into individuals, must depend on secondary processes.

If the reader has followed with assent this somewhat abstruse discussion and elucidation of the identity of Growth and Reproduction, he will have little difficulty in classing the phenomena of Parthenogenesis under the ordinary laws of Reproduction, and removing the peculiar marvel which has hitherto invested those phenomena. Accepting Reproduction as a vital property—an ultimate fact—which appears under various forms of Growth, Gemmation, and Generation, he will admit that there is nothing more marvellous in an animalcule reproducing several millions of animalcules by spontaneous fission, than in a plant being constructed out of several millions of cells, each produced by a spontaneous fission; in each case the marvel is the same—the process the same. It is not more marvellous that an Aphis should produce another Aphis full-formed from its own substance, than that a lobster should out of its own substance replace a broken claw.

The peculiarity of Parthenogenesis which has most attracted and puzzled naturalists is the fact that each generation is *unlike* its parent. In Steenstrup's words: "Generation A produces generation B, which is dissimilar to itself; whilst generation B produces generation C, which is dissimilar to itself, but which returns to the form of generation A." This, on closer scrutiny, becomes very dubious. Agassiz has pointed out the identity in structure of the Medusa and Polype; and although there are formal differences between *these* two animals, in higher animals such differences grow less, and finally disappear. The Aphis produces a larval Aphis, which only differs from its parent in the imperfection of certain organs, and these imperfections are not *constant;* the larva has sometimes wings. The virgin product of the silkworm Moth is every way indistinguishable from the products of fertilised eggs.

What, then, is the theory of Parthenogenesis to which this discussion conducts us? Simply this: The phenomenon is not a *deviation* from the ordinary laws of Reproduction, but a *derivation* from those laws. What *they* are, no one at present can express. The fact that all organic beings are endowed with the property of Reproduction, which manifests itself under the forms of Growth, Gemmation, and Generation, must, for the present at least, be accepted as an ultimate fact, not permitting dispute, not admitting explanation. Whether new individuals or only new parts of individuals, are reproduced, the fundamental process is the same. Whether the animal produce cells which increase as buds, or as eggs, the process is the same. Whether the egg develop under the influence of fertilisation, or without that influence, the process is the same. Whether the union of two cells, followed by continuous fission, be taken as the starting-point, or whether the continuous fissions proceed without any union, everywhere the one law of Reproduction—the fundamental property of Growth—meets us as the ultimate fact, the great terminal mystery; and the simplest form under which this process is known to us is the spontaneous subdivision of a cell. Thus, to borrow Goethe's words—

> " All the forms resemble, yet none is the same as another ;
> Thus the whole of the throng points at a deep-hidden law.
> Points at a sacred riddle." *

The sacred riddle awaits its Œdipus, and probably will for ever remain unanswered.

* *Die Metamorphosen der Pflanzen.*

CHAPTER II.

SUMMER DELIGHTS—MEDUSÆ-HUNTING—NOCTILUCA AND THE PHOSPHORESCENCE OF THE SEA—THE CYDIPPE—VIVISECTIONS—DO THE LOWER ANIMALS FEEL PAIN?—CHANCE-WEED—A NEW POLYPE—A NEW POLYZOON — VITALITY OF MOLLUSCS — VISION OF THE MOLLUSCS—ARE IMAGES FORMED ON THE RETINA?—DESCRIPTION OF THE RETINA IN VERTEBRATES AND INVERTEBRATES — NEW THEORY OF VISION—TACTILE SENSATIONS AND NERVE FILAMENTS—CAN THE MOLLUSCS HEAR?—THE SENSES OF ANIMALS NOT SUPERIOR TO THOSE OF MAN—THE OCEAN-CURRENTS CAUSED BY MOLLUSCS.

THERE are perspiring individuals who love not summer in its sultry splendour. With bubbles on their upper lips, they languidly declare the heat is insupportable. It is not often that our English summers swelter with intolerable heat; and when the blazing sun *does* pour fierce radiance on the land, who have true right to murmur? Only those unhappy victims of civilisation doomed to move along stifling streets, with souls yearning for the far-off woodlands, and the breezy seaboards; or those victims of agricultural necessities who toil amid the shadeless corn. Nobody else. The heat is hot, undoubtedly; but it is beneficent. Nature ripens; life culminates; let no one murmur. I am in a permanent vapour-bath while writing this, yet the temporary discomfort cannot quell my invincible delight in summer: it only gives a more exquisite sense of the evening coolness, and the breezy shade. To walk out under this August sun demands a touch of heroism; yet if we venture out, there is always the refuge of a shady nook behind the rocks, where, sheltered amid the ferns and purple heath, we may recline, and watch the gentle sea lapping the pebbles at our feet. In dreamy mood we "fleet the time carelessly as they did in the Golden Age." A pleasant book beguiles the lazy hour. Murmurous insects sing and labour all around; birds chirp and twitter in their busy joy. These are the psalms of nature, in which the soul finds perennial delight. They sink into our minds with the gentle fall of raindrops in a silent pool, creating many circles. They speak to us of happy days, and chide with their serenity the feverish impatience of our lives.

Then, delicious are our evening rambles, when the birds are ceasing from melodious labour. The lazy toad crawls ungainly from his hole (not despised of us, although the victim of popular prejudice); the timid bat wings its purblind way through the dim air, holding her young one fastened to her breast, and moving with her dear burden less gracefully than her mate; and the numerous goats, browsing on the rocks, are being milked, while their kids are tenderly led home. The sands or the lanes invite us to a meditative stroll, and we ramble on, revolving the various hints, glimpses, hypothetic suggestions, which gather round the facts observed in the morning's labour.

Or, it may be, we step into a boat, and glide softly over the water, skimming its surface with the Medusa-net, to gain fresh material for study. The muslin net, after skimming the surface for two or three minutes, is examined. To the unlearned eye it contains nothing beyond foam-bubbles and stray bits of weed; but we know better. Those bubbles are not all of foam; some of them are exquisite creatures of living crystal; and on reversing the net into the glass jar of seawater, behold! they swim before our delighted eyes as Cydippes, Noctilucæ, and Naked-eyed Medusæ. The *Cydippe* (Plate I., fig. 2) is melon-shaped, with longitudinal bands, on which are transverse rows of very active cilia, not unlike tiny treadmills, and with two long streamers, which follow like the tail of a comet. As we capture these beauties, our boatmen are lost in astonishment. They never " see'd such things afore— that they never did—never in all their lives, long as they've been at sea." Nor can they understand how we distinguish them from the foam-bubbles. Indeed, I cannot myself precisely indicate the characters by which they are recognised; and yet no sooner was there one in the net than it was detected. If the reader desire to learn a simple plan by which he will infallibly detect them, when they escape his rapid eye, let him place his hand underneath the net, where the bubbles are, and the greater opacity of the animals will at once betray them. Then, without loss of time, let him reverse the net into a jar or bucket, and the creatures will float off.

On bringing them home, we place our captives in glass vases, and begin to study them. The *Noctilucæ* are little crystal balls of about the size of a pin's head, which, under the microscope, present the appearance figured in the Frontispiece (fig. 3). The transparence of its structure permits an easy investigation. Not a fibre is to be seen, unless, with de Blainville, we consider the transverse markings of the tail in the light of muscular fibres, a supposition which is very questionable. In the neighbourhood of this tail there is usually a mass of food, or the indigestible remains of food. Not that we are to look for a stomach in this animal—nothing of the kind exists; but in lieu thereof we find, as in Infusoria, a number of *vacuolæ*, or assimilating cavities, which appear and disappear, according to need, formed out of the contractile substance which is seen radiating in filaments all through the substance of the animal, and which M. Quatrefages* likens to the *sarcode* described by Dujardin. In this curious animal, not a trace has been discovered of vessels, nerves, senses, or indeed of any "organs" whatever. It is a mass of animated jelly, with a mobile tail. Its mode of reproduction has been variously expounded, but the observations of Quatrefages and Krohn seem placed beyond a doubt by those recorded in Mr Brightwell's paper,† proving that they multiply by spontaneous subdivision. No one has yet observed anything like reproduction by means of ova.

* *Annales des Sciences Nat.* 1850, p. 231.
† *Microscopical Journal*, No. XX. 1857. p. 185.

To these *Noctilucæ* the sea owes much of that brilliant phosphorescence which at all times has been the marvel of travellers. Place your vase in a darkened room, and strike the glass, or agitate the water, and you will be delighted with the spectacle presented. From every part brilliant sparks appear and disappear, until at length no agitation of the water will produce more; their power is exhausted, as that of the electric eel is exhausted after a few shocks. You want to know the cause of this phosphorescence? Unhappily the point is still *sub judice*. It is only since the beginning of this century that the attention of naturalists has been fixed upon the *Noctilucæ* as sources of the phosphorescence, in all times observed, and in former times attributed to the presence of decaying organic substance, to electricity, to "an absorption of solar light disengaged in the dark," &c.* The extensive and minute investigations of M. Quatrefages led him to the following conclusions:

There are two different kinds of phosphorescence observed in the sea. The first is of very brilliant but isolated sparks, and is due principally to Starfishes, Crustaceans, and Annelids. The second is of a general luminous tint, over which are strewed isolated sparks, and is due to the Noctilucæ. These Noctilucæ have no *special organ* which produces the phosphorescence, as the other animals have; but the light emanates from the whole substance of their bodies. Every

* For the history of these opinions, and other curious details, see the Mémoire of M. QUATREFAGES, *Annales des Sciences*. 1850.

irritant, no matter of what nature, produces this phosphorescence in them. The phenomenon is not, as in insects, one of combustion; but is intimately connected with the contraction, spontaneous or provoked, of their substance. It is independent of all secretion, and it is probable that the *sparks* are due to the rupture and sudden contraction of their sarcodic filaments; while the *steady light* they emit in dying, results from the permanent contraction of this sarcodic substance.

Having satisfied curiosity about the *Noctilucæ*, let us turn once more to the *Cydippes*, which should be placed in the *tallest* jars, because, while the Medusæ keep at the surface, where they swim with successive pants, the Cydippes constantly let themselves drop to the bottom, and rise the next moment in graceful buoyancy, drawing their graceful streamers after them, these streamers elongating as they ascend, until from shrivelled threads they unfold into long and graceful forms, which, on coming into contact with any object, shrink rapidly again into their former shrivelled condition. All this while the locomotive paddles of cilia sway the animal with restless grace—a charming spectacle! After admiring it abundantly, you may commence a closer inspection of the creature's structure, which is sufficiently curious, but need not detain us here, because you may see in any text-book what is known, and I know nothing more than what is there recorded. One remark only need be made: the notion of the streamers (or tentacles) being locomotive organs, as some suppose, is easily disproved; you have only to snip them off, and you will observe the animal moving with the same vigour and grace as before. Nay, if you cut the animal in pieces, each section, provided it has a portion of the ciliated bands, will for days swim about with unabated energy.

The reader, who is of course a lover of animals, and consequently of a sympathetic compassionate nature, will probably feel some repulsion at the quiet way in which he is recommended to snip off the Cydippe's tentacles, and will energetically protest against the cruelty of physiologists who employ vivisection as a means of experiment. It is very true that a grave question has to be answered by the physiologist when, for the sake of science, he inflicts pain. I confess that my susceptibility altogether disqualifies me from witnessing, much more from performing, experiments accompanied with pain. It was a long while before I was able to justify the French and Germans in their wholesale slaughter of puppies, cats, rabbits, and guinea-pigs. Nor can they be justified except by the austere necessities of science. When this is their object, we are wrong to accuse them of cruelty, because cruelty is the indulgence of tyrannous love of power, and their purpose is the grave investigation of truth. Cruel they are not, unless surgery be also cruelty. And in any case the reproach comes with an ill grace from men who torture animals in the way of mere sport, as in hunting, fishing, and the like. I have said thus much in extenuation of vivisections, although, as before intimated, my own organisation renders it impossible for me to witness them in the case of the higher animals. With lower animals the case is altogether different. They feel no pain. If we know anything about them, we know that. You are sceptical? You want to know how it can be *proved* that these animals feel no pain. It is of course impossible for us to say accurately *what* any animal feels; we cannot even know what our fellow-beings feel; we can only approximately guess, interpreting their gestures and cries according to our own experience. Admitting to the full this initial difficulty, we may nevertheless assert that, if it is allowable to make any statement on this point, there are certain capital facts which force the conclusion upon us, that so far from Pain being common to all animals, it is, on the contrary, the consequence of a very high degree of specialisation, and is only met with in animals of complex organisation. It is probable that reptiles have only a very slight capacity for pain, and animals lower than fish none at all.

When we see an animal shrink, struggle, or bite, when we hear it cry or hiss, we naturally interpret these actions as the expressions of pain, because pain calls forth similar actions in us. But there is a fallacy in this interpretation. The movements which in us accompany or succeed the pain are not produced by the organs which feel the pain, even when pain is actually present: they are not produced by pain, but incited by the stimulus pain gives to other organs.

Grief incites the lachrymal organs, but tears flow from vexation, from affliction, from physical pain, or from the effect of an onion on the eyes. Pain incites the vocal organs to a shriek; but we hear persons, unhurt, shriek, when they see others in danger. These illustrations suffice to make clear the difference between movements which *follow* the sensation of pain, and the movements which in themselves indicate it; and enable us to apply the Method of Exclusion, and show that inasmuch as the very same movements are produced by *other* stimuli besides pain, we are not entitled to assume that these movements necessarily indicate pain in all cases. An insect will sometimes continue eating if pinned to the table, and will only struggle to fly away when the food is devoured. "Soft, lubricated, and irritable as is the skin of the naked mollusc," says Professor Owen, "there are not wanting reasons for supposing it to be possessed of a very low degree of true sensibility. Baron Férussac, for example, states that he has seen the terrestrial gasteropods, or slugs, allow their skins to be eaten by others, and in spite of the large wounds thus produced, show no sign of pain."*

It thus becomes evident that shrinking, struggling, crying, &c., are no certain indications of pain. Nay, if we were to accept the shrinking as evidence, we should be forced to admit that the flower feels pain when it shrinks on being touched. The other day I was dissecting a *Solen*, which had already been dead

---

* OWEN: *Lect. on Comparative Anatomy*, p. 551.

eight-and-forty hours, and was beginning to decompose, yet no sooner did the scalpel touch the muscular foot, than that foot shrank, as it would have shrunk in the living animal. Was this pain? Clearly not. It was due to the irritability of the muscular tissue.

Another observation made over the dissecting-table is even more instructive. One of my *Tritons* had been dead some time, and was pinned down on a cork plate by the four paws. I had taken out the heart and lungs, without exciting any obvious contraction, when, on accidentally pricking the tail with the scalpel, I was amazed to see it writhe; repeating the prick, my amazement increased as I saw the whole lower extremities twist and writhe, so as to free the legs from the pins which fastened them to the cork. A bystander would have said that the animal must be suffering pain; yet on pricking the anterior extremities, the ribs, the stomach, and the head, not a trace of sensibility could be detected. Dead the animal assuredly was. He had been dead some hours before I removed his heart, yet sensibility remained apparently as active as ever in the tail; and on examination I observed this sensibility decreased as I ascended from the tail upwards, disappearing altogether midway in the body.

Up to this point, we have done little more than destroy the value of the positive evidence which can be adduced in support of the proposition that all animals feel pain. As regards mere shrinking and struggling, fighting, and crying, we see that the evidence is null. If it should be said that all animals possessing a nervous system must feel pain, because pain belongs to the nervous system, I ask, To what part of that system? We are certain that it does not belong to every part. We have endless nerve-actions incessantly going forward, without a vestige of pain accompanying them. There is no pain in seeing, hearing, thinking, breathing, digesting, &c. If not *every* part of the nervous mechanism, then only some *special* part, or parts, must be credited with sensibility under the form of Pain; and the mere fact of an animal's possessing a nervous system, will aid the argument only when proof is afforded that this system also includes the special part or parts endowed with sensibility to Pain.*

As far as I can see into this obscure question, Pain is only a *specialisation of that Sensibility which is common to all animals*. It is a specialisation resulting from a high degree of differentiation of the nervous system, consequently found only in the more complex animals, and in them increasing as we ascend the scale. Out of a primordial basis of Sensibility (one of the vital properties—an ultimate fact, therefore), various special forms are developed. In the ascending series we have first reflex action, we have next the organic sensations, then the special sensations of seeing, hearing, tasting, smelling, touching; we have, further, the sensations of shivering, tickling, fatigue, hunger, thirst,

* In the *Proceedings of the Liverpool Literary and Philosophical Society*, No. IV. (1848), will be found an interesting essay by Dr Inman, entitled "On the Non-existence of Pain in the Lower Animals," in which many curious facts are collected.

which, although not painful in themselves, may easily pass into pain. Finally, we have a specific form of Sensibility capable of being excited by a great variety of stimuli in great variety of degrees: and this is Pain; which appears to exist in all the higher animals, though in a feebler degree than in man. Even among men the difference of susceptibility is very remarkable. It is much less in savages than in highly-civilised men, as it seems also to be less in wild animals than in domesticated, especially petted, animals; less in men leading an active out-of-door life than in those leading a sedentary intellectual life; less in women than in men; less in persons of lymphatic than in persons of nervous temperaments. To one man the scratch which is a trifle scarcely noticed, is to another an obtrusive pain; the one will not even tie his handkerchief over the wound, so little does it press upon his sensibility; the other is pale, and must have the wound dressed.

It is because men habitually confound Sensibility with Pain—the general with the particular—that so many disputes continue respecting the sensibility of certain parts of the nervous mechanism; for instance, the disputes as to whether the Sympathetic system is also a Sensitive system. But no correct understanding of the nervous system can be arrived at until more rigorous language is adopted, and we learn to designate all nerve-actions by the one general property of Sensibility, and to discriminate between this general property and its special manifestations. Pain is, I believe, a special form of this general Sensibility; and although science has not yet detected the special condition whereby stimuli are transformed into sensations of pain, there can be little doubt that such a condition exists, and none at all, in my mind, that the lower animals have it not; and this conviction keeps me perfectly calm in performing experiments on marine animals: a very desirable result, seeing that, without experiment, our observations would carry us but little way.

But let us turn to another subject. In describing the various methods of search for animals, it has been assumed that a tolerable conception exists as to the appearance of the thing sought. We may also count on "chance-seeking." We never know all that we have captured until some days afterwards. Repeated examinations of our vases and bottles with a lens, enable us to detect many a curious novelty which was unsuspected among the weed, and has now emerged. It is, therefore, a good plan always to bring home some "chance-weed," especially if it have a root; the red weeds being the most advantageous. This is placed in sea water for a day or two, and carefully examined from time to time; something is tolerably certain to be found thereon. One day, going over the contents of a bottle with a lens, I was struck by the curious appearance of some Sertularian Polype, round which minute grains of sand seemed to be clustered, but all equidistant from the Polype, and not visibly attached to it. On removing it to the stage of the microscope, these supposed grains of sand proved to be the cups of a tiny

z

Polype, in aspect closely resembling *Tubularia indivisa*, growing parasitically on the Sertularia. Proceeding to identify it, I found the species to be one hitherto undescribed; and I propose to name it *Tubularia parasitica*, if no one has been before me. On another occasion I saw, with the naked eye, a polype-like creature attached to the side of the glass, with its tentacles expanded; the lens showed it to be a Polyzoon, much resembling the *Alcyonidium hirsutum*. It was single, however; and on other parts of the glass were eight other specimens, all solitary. This was in itself noticeable, because, as the name Polyzoon imports, these animals live in colonies. Under the microscope, a new fact presented itself: the animal was enclosed in an oblong bivalve-shell, which seemed permanently open on one side, and open at the summit to give passage to the crown of tentacles. Imagine a shell like that of a mussel gaping open, within which is a quinine bottle, the broad neck protruding, and you will form a tolerable idea of the general aspect of this animal when the tentacles are withdrawn. I believe this to be a new genus, and also to have an interest beyond novelty, because furnishing another decisive argument in favour of the molluscan nature of the Polyzoa—a point still disputed among naturalists.* The existence of a bivalve-shell is very important; and I took pains to convince myself that it was really a shell, and not a membranous envelope having the aspect of a shell:

---

* The recent publication, by the Ray Society, of Professor ALLMAN'S monograph on *Freshwater Polyzoa*, must for ever settle this dispute.

submitting the animal to decomposition, I found the shell remain behind intact.

Apropos of Molluscs: their powers of endurance are very remarkable. Having noticed that they live out of their native element, the water, for a considerable time, being often left bare on the rocks by receding tides, I thought of testing their powers in this way. Accordingly, a Cockle was placed on my work-table, out of all reach of damp, in a room where a fire was constantly burning. This was on the 10th of April; not until the 21st was the cockle dead. A small fish (*Ophidium*) under similar circumstances died in seven hours. Whence this remarkable difference in two gill-breathing animals? A question easily asked, but not easily answered.

It is true that both animals are aquatic, and both breathe by gills; but when we come to understand the complex mechanism of respiration, we see various special differences between the two organisms. Let us begin with that of the fish. M. Flourens* has shown that the weight of the soft leaflets composing the fish's gill differs but slightly from that of water; so that when the animal is in water the slightest force suffices to float and separate them, by which means the water bathes their surfaces, and the exchange of gases takes place. But no sooner is the fish brought out of the water than the difference between the weight of its gills and that of the atmosphere, immediately causes a col-

* *Expériences sur le Mécanisme de la Resp. des Poissons;* in the *Annales de Soc. Nat.*, 1830, p. 5.

lapse of the former: the leaflets, instead of floating free in the air, are pressed together, so that only the external surface of the two outer leaflets are in contact with the air, and this is obviously too small a surface to suffice for the whole aeration of the blood; and the fish dies of asphyxia. Add to this cause, the rapid dessication which ensues on exposure to the air, and which we know is an obstacle to respiration.

The Mollusc—our Cockle for instance—is somewhat differently provided. It is true the molluscan gills are formed of leaflets heavier than the air; but when we take him from the water he closes his shell, and in that shell a reasonable supply of water remains. But this is not his chief safeguard. A constant exudation from the surface keeps the gills *moist*, and this moisture permits the exchange of gases, on which respiration depends. It is *this* cause which enables the land-crabs to live in the atmosphere, although their gills are formed on the same plan as those of the marine crabs. Milne Edwards has shown that a special reservoir exists which preserves the humidity of their gills.* In those Molluscs which have no supply of water in their shells to keep the branchiæ floating, there is always a constant moisture to keep them fit for respiration; and although the respiration must necessarily be feebler under such circumstances, yet we must remember that the vital changes are not so rapid in its lethargic and comparatively simple organism as in that of the fish.

The molluscs do not rank high in the scale of intelli-

* See his *Leçons sur la Phys. et l'Anat. Comparée*, 1857, vol. i. p. 519.

gence, yet even the Oyster seems to be educable to a small degree. Milne Edwards relates, that in the great oyster establishments on the coasts of Calvados, he learned that the merchants teach these succulent molluscs to keep their shells closed when out of the water, by which means they retain the water in their shells, keep their gills moist, and arrive lively in Paris. The process is this: No sooner is an oyster taken from the sea than it closes its shells, and opens them only after a certain time—from "fatigue," it is said, but more probably because the shock it received, and which caused its muscles to contract, has passed away. The men take advantage of this to exercise the oysters, and make them accustomed to be out of water, by removing them daily into the atmosphere, and leaving them there for longer and longer periods. This has the desired effect; the well-educated mollusc keeps his shell closed for many consecutive hours, and as long as the shell is closed his gills are kept moist.

The ten days of my Cockle sink into insignificance beside the astonishing facts on record. In Mr Woodward's valuable *Manual of the Mollusca*, we read: "The fresh-water molluscs of cold climates bury themselves during winter in the mud of their ponds and rivers; and the land-snails hide themselves in the ground, or beneath the moss and dead leaves. In warm climates they become torpid during the hottest and driest part of the year. Those genera and species which are most subject to this summer sleep are remarkable for their tenacity of life, and numerous instances have been recorded of their importation from distant countries in a living state. In June 1850, a living pond-mussel was sent to Mr Gray from Australia, which had been more than a year out of water. The pond-snails have been found alive in logs of mahogany from Honduras; and M. Cailland carried some from Egypt to Paris packed in sawdust. Indeed, it is not easy to ascertain the limit of their endurance; for Mr Laidlay, having placed a number in a drawer for this purpose, found them alive after *five years*, although in the warm climate of Calcutta. Mr Wollaston has told us that specimens of two Madeira snails survived a fast imprisonment in pill-boxes of two years and a half. But the most interesting example of resuscitation occurred to a specimen of the desert snail from Egypt, chronicled by Dr Baird. This individual was fixed to a tablet in the British Museum on the 25th March 1846; and on the 7th March 1850 it was observed that he must have come out of his shell in the interval (as the paper had been discoloured, apparently in his attempt to get away), but finding escape impossible, had again retired, closing his aperture with the usual glistening film; this led to his immersion in tepid water and marvellous recovery. He is now (March 13, 1850) alive and flourishing, and has sat for his portrait."

The Molluscs, like the heathen idols, have eyes for the most part, yet see not; organs of hearing, yet hear not; nevertheless, unlike the heathen idols, they are endowed with these organs for no " make-believe," but for specific purposes. A function there must be, and doubtless a good one; but we speak with large latitude of anthropomorphism when we speak of the "vision" of these animals. Molluscan vision is not human vision; nor in accurate language is it vision at all: it is not *seeing*, but *feeling;* it is not a perception of objects, but a sensation of light and darkness. This does not apply to the Cephalopoda, in which vision seems to be as perfect as in Fishes; nor, on the other hand, does it apply to those Bivalves which have no eyes at all, not even "eye-spees." The word Mollusc embraces a vast variety; and, by way of limitation, the reader must understand that the following remarks are confined to those genera which I have directly studied for the purpose — Doris, Eolis, Pleurobranchus, and Aplysia. In the three first genera the eyes are *underneath* the skin and muscles, and rest on the brain (œsophageal ganglia), attached thereto by a microscopic nerve. There is no aperture in the skin, as there is in ours, through which the rays of light may fall directly on the eye; so that in spite of pigment, lens, and nerve — the essential parts of a visual organ — vision is utterly impossible; as you may convince yourself even with your own admirable eyes, if the lids are obstinately closed over them. I am aware that clairvoyants of the strictly unveracious species, profess to see with their eyes closed; but our simpler Molluscs have no such pretensions; they have not yet given in to the clairvoyant mania, and are content to submit to those laws of physics which regulate phenomena with the same unerring consistency in the world of Naked-gills as in that of Clothed Noodles.

A first requisite in vision is surely the formation of an image; and how can this image be formed when the rays have to pass through the skin and muscles covering the eyes? A second requisite is a special ganglion, or centre of sensation; and even this is wanting in many cases. In *Pleurobranchus* and *Aplysia* I find the optic nerve arising from the ganglion which supplies the antennæ; and Leydig says the *Doris lugubris* has its small eyes resting immediately on the brain.* Nevertheless, although these eyes are incompetent to vision, they represent the early stages of that marvellous and complex function; they are special organs for the reception of luminous influence, enabling the animal to distinguish light from darkness, not only in the general way, like a blind man conscious of a change of temperature in passing from sunlight into shade, but also in the special way of minute local variations, such as are caused by the shadows of near objects.

I remember once being seated with a philosophic friend, and much bored by the presence of a morning caller—a large white-waistcoated man, "such an ass, and so respectable!" stiff with ignorance and haughtiness: the kind of man who seems afraid of lowering his eyebrow lest it should crease his cravat. He droned away about "the house" and Lady Jane, about his

---

* LEYDIG: *Histologie d. Mensch. u. Thiere*, 1857, p. 249. I have also observed this in a species of Doris of which the name is unknown to me. In general the *Dorida* have minute optic ganglia.

tenants, and what he had said on several occasions, till my patience was exhausted; and thinking nothing more likely to hasten his departure than a touch of Transcendental Anatomy, I turned to my friend, and, as if resuming the thread of our conversation, remarked, "Yes, it is singular to think of the eye being nothing more than a tactile organ." Whereupon White-waistcoat precipitately retreated. He would not wait to hear the development of that mad proposition; yet, had he waited, he might have learned that the eye *is* a tactile organ, and that what we call vision is a combination of the sensations of touch, and of temperature of a specific kind.

The common notion is, that objects are reflected as images on the retina, and thence, as images, transmitted to the brain. But *nous avons changé tout cela*. I have serious doubts whether an "image" is formed on the retina at all; and the strongest conviction that no image is transmitted to the brain; on the contrary, the thing transmitted is a *sensation*, or group of sensations, *excited* by what is called the "image." The wave of light is translated into a nerve-stimulus, the impression excites a sensation; but the sensation is due to the specific centre, not to the specific stimulus of light; as we know by the fact that any *other* stimulus, such as pressure or electricity, is translated into a precisely *similar* sensation. So that even if we suppose an image to be formed on the retina, as it is formed in a camera-obscura, *it* will not be transmitted to the brain, but it will excite the specific sensations of which the optic centre is alone capable, and *these* will be transmitted.*

When, a little while ago, I said that the formation of an image was a primary requisite in vision, I meant that unless the rays from an object converged into an image on a proper surface, no distinct perception of that object could result. The reader will not, therefore, suppose that, in throwing doubt on the notion of images being formed on the retina, as they are formed on the camera-obscura, any attempt will be made to overthrow the optical principles minutely established by philosophers. A brief description of the retina and its connections will enable us to argue this point at our ease.†

The retina is not, as commonly supposed, simply an expansion of the optic nerve—if by that be meant a purely fibrous layer; it is more accurately described as a membraniform ganglion. After entering the eye, the nerve expands, and lines the inner surface with a layer of fibres; but beneath this layer is one of cells, not distinguishable from those of the brain, and beneath that, one of granules; beneath this layer, again, is another of perpendicular rods and cones, known as the "membrane of Jacob." So that we have four distinct layers, very dissimilar in structure, and of course very different from the optic nerve, which is simply fibrous. Instead of regarding the

---

* "Light and colour are *actions of the retina*, and of its nervous prolongations to the brain."—MUELLER: *Physiology*, Eng. Trans., p. 1162.

† The student should carefully read KÖLLIKER, *Handbuch der Gewebelehre*, and H. MÜLLER, *Anatomisch-physiologische Untersuchungen über die Retina bei Menschen u. Wirbelthieren*. 1856.

retina as composed of layers, however, modern investigators, following Kölliker and Heinrich Müller, are generally agreed in considering that the fibres of the optic nerve pass *radially* through the retina: thus from the fibres a thread passes downwards till it meets a cell of the vesicular layer, which in turn is in connection with a granule of the granular layer, which terminates in a cone and rod; these latter forming the *real termination of the optic fibre in the pigment layer of the choroid coat.*—(See Plate III., fig. 4.) It is now universally held that the rods and cones are the percipients of light, which they communicate to the cells of the vesicular layer, thence to the optic fibres, and thence to the optic ganglion. The point to be borne in mind in this description is that the *sensitive part of the retina is not the surface on which the light immediately falls, but the surface which is in contact with the black pigment.*

In a parenthesis I may add, that one of the Dorpat school[*] has considerably disturbed the harmony which existed on the subject of the retina, by the publication of a series of researches, which led him to the conviction that only the optic fibres of the retina are of nervous structure, the rest being formed of "connective" tissue. Whatever may be the issue of the quarrel thus raised, it will not affect the points to which our argument will be directed; indeed, Funke[†] already suggests that, inasmuch as the function of

---

[*] BLESSIG: *De Retinæ Structura*: 1855. See an abstract in CANSTATT's *Jahresbericht*: 1855.

[†] *Lehrbuch der Physiologie*, 711.

the rods and cones is one to which nervous tissue is confessedly incompetent—namely, the transformation of the wave of light into that molecular process which takes place in the conduction of the impression—we may readily admit that their structure is different.

From what has been already said, it will be easy to prove that no images can be formed on the surface of the retina. In the first place, the retina, during life, is as transparent as glass. The rays of light must therefore pass *through* it, and enter the pigment layer, which, being perfectly black, absorbs all rays. Further, it has been proved that the optic fibres are *totally insensible* to light. There is a blind spot in each eye. Would you know the peculiarity of that spot? It is where the optic nerve enters, and where, consequently, nothing but nerve-fibres exist. There is also a spot in each eye where the sensitiveness to light is at its maximum. Would you know the peculiarity of that spot? It is a mass of cells, without a continuous surface-layer of fibres. After proving that the fibres are insensible to light, and that no image is formed where the fibres alone exist, we are called upon to show that some apparatus exists for the reception of these rays of light out of which the necessary images are formed; and to Professor Draper we must turn for the best hypothesis to aid us.

Franklin, he reminds us, placed variously-coloured pieces of cloth in the sunlight on the snow. They were so arranged that the rays should fall on them equally.

After a certain period he examined them, and found that the black cloth had melted its way deeply into the snow, the yellow to a less depth, and the white scarcely at all. The conclusion which he drew has since been abundantly confirmed; namely, that surfaces become warm in exact proportion to the depth of their tint, because the darker the surface the greater the amount of rays absorbed. A black surface, absorbing all rays, becomes the hottest. This principle Professor Draper invokes in his examination of the eye. The pigment layer is, he maintains, the real optical screen on which the images are formed: "The arguments against the retina, both optical and anatomical, are perfectly unanswerable. During life it is a transparent medium, as incapable of receiving an image as a sheet of clear glass, or the atmospheric air itself; and, as will be presently found, its sensory surface is its exterior one— that is, the one nearest the choroid coat. But the black pigment, from its perfect opacity, not only completely absorbs the rays of light, turning them, if such a phrase may be used, into heat, no matter how faint they may be, but also discharges the well-known duty of darkening the interior of the eye. Perfection of vision requires that the images should form on a mathematical superficies, and not in the midst of a transparent medium. The black pigment satisfies that condition, the retina does not."*

Now comes the difficulty. If the retina is insensible to the light which passes *through* it, it will be equally

---

* DRAPER: *Human Physiology*, p. 387.

insensible to the light which, according to some physiologists, is reflected from the pigment layer.

On the other hand, although the pigment layer is capable of absorbing light, we cannot suppose it also sensitive to light. How, then, is the luminous sensation produced? Professor Draper shall again furnish us with an answer:—"The primary effect of rays of light upon the black pigment is to raise its temperature, and this to a degree which is in relation to their intensity and intrinsic colour; light which is of a yellow tint exerting, as has been said, the most energetic action, and rays which correspond to the extreme red and extreme violet, the feeblest. The varied images of external objects which are thus painted upon the black pigment, raise its temperature in becoming extinguished, and that in the order of their brilliancy and colour. . . . *In this local disturbance of temperature the act of vision commences;* this doctrine being in perfect harmony with the anatomical structure of the retina, the posterior surface of which is its sensory surface, and not the anterior, as it ought to be, if the explanation usually given of the nature of vision is correct; and, therefore, as when we pass the tip of the finger over the surfaces of bodies, and recognise cold and warm spaces thereupon, the same process occurs with infinitely more delicacy in the eye. The club-shaped particles of Jacob's membrane are truly tactile organs, which communicate to the sensory surface of the retina the condition of temperature of the black pigment."

It is worth remarking that the analogy in structure

between the retina and the recently-discovered tactile corpuscles is very close.* Professor Draper further insists on the fact that all photographic effects result from high temperature: "The impinging of a ray of light on a point raises the temperature of that point to the same degree as that possessed by the source from which the ray comes, but an immediate descent takes place through conduction to the neighbouring particles. This conducted heat, by reason of its indefinitely lower intensity, ceases to have any chemical effect, and hence photographic images are perfectly sharp on their edges. It may be demonstrated that the same thing takes place in vision, and in this respect it might almost be said that vision is a photographic effect, the receiving surface being a mathematical superficies, acting under the preceding condition. All objects will therefore be definite and sharply defined upon it, nor can there be anything like lateral spreading. If vision took place in the retina as a receiving medium, all objects would be nebulous on the edges."

To explain the process by which the change of temperature in the pigment becomes a luminous sensation will not be difficult, if, remembering that the luminous sensation is one not depending on the specific stimulus of light, but on the specific nature of the optic centre, we follow this change in its passage from the pigment to the rods and cones of Jacob's membrane, which it first affects; these are in direct connection with the

* See LEYDIG, *Histologie:* and FUNKE, *Physiologie*, where diagrams are given.

ganglionic nerve-cells, in which we may suppose the nervous impression to be excited; this impression is thence transmitted by means of the optic fibres to the optic ganglion, and there it becomes a sensation. This is hypothetical, I admit; but it is the only hypothesis which can agree with the present condition of our anatomical knowledge. Funke has a good illustration. The wave of light, he says, can no more excite the optic nerve *directly*, than the pressure of a finger on the air, or the walls of the organ-pipes can excite musical notes. The finger produces a tone by pressing on the keys; each particular key that is pressed brings forth a corresponding tone as the air enters the pipe. In this illustration the optic fibres are as the organ-pipes, the rods and cones of Jacob's membrane as the keys, and the wave of light as the wave of air.

The most convincing argument against the retina as the receiving screen of images, and in favour of the pigment layer, is, in my opinion, to be found in the eyes of the Invertebrata, where the *pigment is in front of the retina*, instead of behind it, as in the Vertebrata. I have examined this point with great care, and the result is, that, although in crabs and insects, for instance, radial fibres in connection with the retina pass through the pigment, and are consequently exposed to the light, yet in every case the vesicular and granular layers and the optic fibres are *beneath* the pigment. In the eye of the Cephalopoda this position of the pigment has long been a puzzle, and Professor Owen says that it must doubtless be "perforated by the retinal papillæ, or otherwise a perception of light must take place, in a manner incompatible with our knowledge of the ordinary mode in which the retina is affected by luminous rays."* True, but the ordinary mode of conceiving the process, we have just seen to be untenable. When Von Siebold says that the "mysterious phenomenon rests only on an imperfect knowledge of the structure of the organ,"† he seems to me to forget that the phenomenon is by no means peculiar to the Cephalopoda, but is characteristic of the Invertebrata generally. What, for instance, is the simplest form of an eye, disregarding those hypothetical "eye-specs" which have been noticed in Infusoria? It is that of a pigment spot *on* a ganglion, or a nervous expansion. Ascending higher in the scale, and reaching even the complex structure of the crab's eye, what do we find but a pigment layer *covering* the retina? If certain processes do pass through the pigment from the retina, it is very questionable whether these are nervous in structure, and, if nervous, they are still only conduct-

---

* OWEN: *Lectures on Comp. Anatomy*, p. 585. But he confesses not to have seen such perforations. I have tried in vain to discover any. In *front* of the retina there is a delicate membrane, but it has none of the characteristics of a nervous tissue, nor have I been able to trace any communication between it and the retina, *through* the pigment. Even should such a communication exist, the ordinary theory of vision would derive little support from it.

† VON SIEBOLD: *Comp. Anatomy*, p. 284. Very imperfect our knowledge is; although on what evidence Professor Rymer Jones (*Animal Kingdom*, p. 591) denies the existence of the choroid, I know not. I have not only seen it repeatedly, but have made a preparation which exhibits it very clearly.

ing-threads, insensible to the direct influence of light. They are held to be analogous to the rods and cones of the vertebrate retina, which, as we have seen, receive their stimulus from changes in the pigment, *not* directly from the light. It is thus, as Leydig says, "in the Vertebrata the rods form the outermost layer of the retina; in the Invertebrata they form the innermost. Herewith is connected the fact, which at first seems so surprising, that the choroideal pigment lies in front of the retina, therefore the contrary of what occurs in Vertebrata."* In the blind Crustacea no pigment is present; and in Albinos, in whom the pigment is of lighter colour, vision is imperfect. If we remember that, according to the hypothesis, light only affects the retina after changing the temperature of the pigment, which change is communicated to the rods and cones, and thence to the vesicular layer, there will be nothing paradoxical in this inverse arrangement of the retina in Invertebrata; in both, the process is essentially the same, and the mere difference of position is not more than the difference of the chain of ganglia, which in the Vertebrata is dorsal, and in the Invertebrata ventral.†

Returning from this digression, and its surprises, to

---

* LEYDIG: *Histologie*, p. 253.

† Lest it should be supposed I have overlooked it, I will notice one serious difficulty in the way of the hypothesis just expounded, namely, the existence in some animals of a strongly reflecting membrane—the *tapetum*—between the retina and pigment layer. I do not at all understand the way in which this affects vision, either on the old or new hypothesis.

the eyes of our Nudibranchs, we can have little doubt that their vision is simply the perception of light and darkness. The changes of temperature produced by the absorption of the rays in their pigment, cannot be elevated into the perception of an image, because the optical conditions for the formation of an image are absent: an indefinite sensation, resulting from change of temperature, is all that they can perceive. Nay, even were their eyes constructed so as to form optical images, there is little doubt that vision, in our human sense, would still fail them, owing to the absence of the necessary combination of tactile sensations with sensations of light. We see very much by the aid of our fingers.

Apropos of tactile sensations, are those anatomists who assume the existence of invisible nerves in parts of the skin which, although revealing no nerve to the eye, seem to reveal it to the mind by the manifestation of sensibility, warranted in such an assumption? Kölliker has shown that there is no portion of the skin, however minute, which is not sensitive. But does this prove that every point must be supplied with a nerve? Admitting that sensibility resides *only* in nerve-tissue, I think another explanation will do away with such an assumption. It is unnecessary that a nerve-fibre should be directly pressed upon at the immediate point of contact of the needle and the skin. The sensation will equally result if the pressure be communicated at some distance *from* the point of contact. Strictly speaking, this is always the case when the cuticle is not pierced. The needle presses on the cuticle, and the pressure is communicated from the cuticle to the nerve; and it is evident that this pressure may be lateral as well as perpendicular. If a nerve be within the range of this lateral pressure, it will be affected; and although those parts which are liberally supplied with nerves are necessarily more sensitive than others, because more filaments come within the range of lateral pressure, yet no part of the skin is insensible, because no part is without the range of a nerve.

Having ascertained that our Molluscs cannot see, we have now to inquire whether they can hear. As in the former case, the answer must depend on what is meant by "hearing." If every sensation of light and darkness is to be called sight, and every sensation of sound is to be called hearing, our friends certainly both see and hear—as blind men see, and deaf men hear. Let us examine the organ in a *Doris* or *Pleurobranchus:* instead of the complex structure found in higher animals, we find a microscopic vesicle containing pebbles suspended in liquid. In the *Doris* this vesicle has no nerve, but lies upon the cerebroid ganglion, immediately behind the optic ganglion. Nor have I, in a dozen dissections, been able to detect a nerve in the *Pleurobranchus*, although Krohn describes one in the sub-genus *Pleurobranchæa*. At any rate, embryology proves the nerve to be a subsequent addition, since in the embryos of all the Nudibranchs the ear is a simple vesicle containing a single otolithe, with neither nerve nor ganglionic attachment. The mention of embryological indications reminds me that Von Siebold has shown the close analogy which exists between the permanent organ of hearing in the gasteropod Molluscs, and the transitory form of that organ in the embryo of the fish.

With such an organ, a mere bag of pebbles in liquid, only a slight degree of that exquisite sense, known by us as Hearing, can be claimed by the interesting animal which naturalists are fond of styling " the humble Mollusc." *I* never detected any humility in Molluscs; and if they seem humble in the eyes of haughty ignorance, a little knowledge of their structure will soon remove that misconception. It is true, they give no dinners, and are perfectly regardless of the higher circles; they trouble themselves very little about any of the " great movements ;" they do nothing for the " Progress of the Species;" leave the Jews unconverted; have no views on the " Ballot ;" and are utterly insensible to the advantages of " Marriage with a Deceased Wife's Sister." But they have their little world, and are as perfectly constructed for it as we are, who condescend to notice and patronisingly admire them.* In that world they do not need what we need. They hear nothing of the marvellous inflections of speech, the tremulous tenderness of affection, the harsh

---

* " Les mollusques sont les pauvres et les affligés parmi les êtres de la création," says Virey, who originated the principle of Cuvier's classification, but who was talking at random when he thus spoke. In creation there is neither high nor low; there are only complex and simple organisations, one as perfect as the other.

trumpet-tones of strife, the musical intonations of mirth. They cannot hear the prattle of children's voices, which sends such thrills along our nerves; nor can they hear the untiring eloquence of a vexed virago, which also sends thrills, not of so pleasant a nature. Deafer than the deafest adder will they remain, charm we never so wisely. Equally insensible must they be to music. Beethoven's melodious thunder, Handel's choral might, Mozart's tender grace, Bellini's languorous sweetness, are even more lost on them than on the lymphatic dowagers in the grand tier, who chatter audibly of guipure and the last drawing-room, while Grisi's impassioned expression, and Mario's *cantabile*, are entrancing the rest of the audience. The Mollusc can only perceive noises. Sounds are by us separately recognisable in their intensity, their pitch (or note), and their quality. The Mollusc only recognises intensity—loudness. A wave of sound agitates the otolithes in his ear, and their agitation communicates to the ganglion a sensation of sound, loud in proportion to the agitation.

Had we no other evidence, this would suffice to show the error of the vulgar conception of hearing. Sound is not produced by waves of air striking the drum, these waves being thence transmitted along the auditory nerve to the brain; but the waves agitate the sensory apparatus, which in its turn acts upon the Sensational Centre. That is why sounds are heard with painful distinctness when the sensory apparatus is affected by other stimuli besides the pulsations of waves of air.

Few subjects are of greater interest to the philosophic mind than the gradual complication of the organ of hearing, with, of course, its proportional complication of function, in the animal series. Even in human beings we see differences only less considerable than those which exist between man and animals. The ear of one man is utterly incapable of appreciating those delicate intervals and harmonic combinations which give to another exquisite delight. The bird,

"Singing of summer in full-throated ease,"

is insensible to music, and probably distinguishes nothing in speech except the loudness of the tone. And this fact may lead us to question whether the general notion, so often insisted on, of the superiority in the senses of animals over those of man, is not a fallacy. It is quite true that a bird sees distinctly at greater distances than a man; but can it see such delicate *nuances* of colour? A dog perceives some odours to which we are insensible; but in the immense varieties of odours we are capable of perceiving, our superiority is manifest. In hearing, animals are demonstrably inferior. Some of them may be as susceptible to certain sounds, but none are susceptible to the immense variety of sounds distinguishable by our ears.

Before quitting our Molluscs, let us for a moment consider the shells with which the vast majority are furnished, and with which all are furnished in their embryonic state. I do not mean that we should lose ourselves in the varieties of a conchologist's collection, nor that we should inquire minutely into the structure of the shell and its mode of growth; but that we should pause to consider its relation to the great forces of the universe. You may possibly look upon that phrase as mere rhetoric; but it is of strictly scientific sobriety; and you will admit it to be so, on learning that the mighty ocean-currents mainly depend on this said mollusc-shell. Strange, yet true. Were there no secreting animals in the sea capable of removing from the water its surplus lime, the stormy winds might agitate its surface, and rouse its waves like troops of roaring lions shaking back their manes of spray; but there would be no strong currents with beneficent effect; and in a little while the ocean would become a huge salt-lake.

Let us rest from our hot hammering, and painful stooping under ledges, and let us enjoy a few minutes' repose on this reef, solitary amid the waves, and distant from the shore. Pleasant the breeze, pleasant the gentle cadence of the water at our feet, pleasant the sight of that snowy mass of cloud which lazily rolls landwards. It rose from the surface of this brilliant, buoyant, volitant sea in airy bubbles of vapour, and is now travelling towards those green cornfields over which the lark is poised in melody. If the cloud should there meet a current of cold air, it will drop gently down as rain. This rain will make its way through the earth to rivulets and rivers, till it finally returns once more to the parent-bed of ocean; but on its way it will have washed with it various salts, which it will dissolve and carry to the sea, thus adding to the already saturated sea water an amount of solid matter such as would impede its flow, were there no provision ready to restore the equilibrium. For observe, the rain-cloud, as it rose by evaporation from the sea, left behind it all the salts which it contained, and these would make the rest of the water denser; but now the rain-cloud returns laden with as much salt as it originally had, and the very fluidity of the sea is in peril, for evaporation is incessantly going on, and rivers are incessantly returning laden with lime. What becomes of this excess of lime? Polypes and Molluscs, Crustacea and Fish, but mainly the two former, clutch hold of it, wring it from the water, and mould it into habitations for themselves. It is thus that vast coral islands and oyster-beds are formed. The sea is a great lime-quarry; but the lime is arranged in beautiful forms, and subserves a great organic end. Not only are animals thus furnished with houses and solid structures, but the water, relieved of its excess, is enabled to flow in mighty currents. This is the theory propounded by Lieutenant Maury in his fascinating book.* Assuming the waters of the sea to be in a state of perfect equilibrium, the animals would, by their secretion of salts from it, produce currents: "The Mollusc, abstracting the solid matters, has by that act destroyed the equilibrium of the whole ocean, for the specific gravity of that portion of water from which this solid matter has been abstracted is altered. Having lost a portion of its solid contents, it has become specifically lighter

* MAURY: *Physical Geography of the Sea*, p. 167.

than it was before; it must, therefore, give place to the pressure which the heavier water exerts to push it aside, and occupy its place; and it must consequently travel about and mingle with the waters of the other parts of the ocean, until its proportion of solid matter is returned to it. . . . . The sea-breeze plays upon the surface; it converts only fresh water into vapour, and leaves the solid matter behind. The surface thus becomes specifically heavier, and sinks. On the other hand, the little marine architect below, as he works upon his coral edifice at the bottom, abstracts from the water there a portion of its solid contents; it therefore becomes specifically lighter, and up it goes, ascending to the top with increased velocity to take the place of the descending column, which, by the action of the winds, has been sent down loaded with fresh food and materials for the busy little mason in the depths below."

Was I not justified in saying that the Mollusc was deeply interesting in its relations to the great forces of the Universe? Does not this one example show how the great Whole is indissolubly connected with its minutest parts? The simple germination of a lichen is, if we apprehend it rightly, directly linked with the grandest astronomical phenomena; nor could even an infusory animalcule be annihilated without altering the equilibrium of the universe.

> "Nothing in this world is single :
> All things by a law divine
> In one another's being mingle."

Plato had some dim forecast of this when he taught that the world was a huge animal;* and others, since Plato, when they conceived the universe to be the manifestation of some transcendent Life, with which each separate individual life was related, as parts are to the whole.

* PLATO: *Philebus*, p. 170, ed. Bekker.

CHAPTER III.

HOW TO CATCH RAZOR-FISH—THE CORKSCREW CORALLINE—DANGER OF
A PRIORI VIEWS IN ZOOLOGY—EXAMINATION OF THE NERVOUS
SYSTEM OF MOLLUSCS—DOUBTS RESPECTING CURRENT DOCTRINES
OF NERVE-PHYSIOLOGY—ABSENCE OF NERVE-FIBRES—ORIGIN OF
SENSIBILITY—SENSIBILITY IN THE ABSENCE OF NERVES—ON THE
RELATION BETWEEN ORGAN AND FUNCTION—CONCLUSION.

THE juvenile naturalist is often instructed in the facile art of catching birds by first dropping a pinch of salt upon their tails. Excellent as this plan seems, it has never proved perfectly successful, owing to a trifling initial difficulty. If, in your halcyon days, you have ever made the attempt, what will your thoughts be on hearing that an intense philosopher, like the present writer, did actually imitate that attempt in capturing the Razor-fish (*Solen*); positively carrying a paper of salt, in the firm confidence of dropping some grains on the tail of that retiring mollusc? Nay, what will you think on hearing that this was not only attempted in all seriousness, but in all seriousness succeeded?

The shells of the Solen lie scattered on the sands of all our bays, and are familiar to every sea-side visitor.

They are not unlike razor-handles, and as every marine animal is called a fish, the Solen is hence named Razor-fish, although, in truth, a bivalve mollusc. Few persons, except naturalists, have seen the animal alive. He bores a hole many feet in the sand, and there passes his days, like the *Pholas* in the rock, never coming to the surface, and boring out of all reach when disturbed. The difficulty of getting at him is obvious, and may lead you to be sceptical of the salting plan. Surely, you will argue, the initial difficulty which discourages the ornithological ardour of the boy must equally frustrate the philosopher; since, if the Solen will allow himself to be approached near enough to have his tail salted, he may be taken *without* this saline preliminary. Thus reasons the reader, not without astuteness; yet, like many other reasoners, he will find that *a priori* deductions, however elegant, frequently pass over the head of Fact. Instead, therefore, of arguing what must be, let him come with me, and see what is. He will find that the Solen must first be salted, ere he suffer himself to be approached.

It is a hot, quiet afternoon. The tide is out, and a wide sweep of sand lies before us. We are armed with thin iron rods, each barbed at the end like a harpoon; we add thereto a paper of salt, basket, and jar. Over the yielding sand we pass, until we approach low-water mark, and then we begin peering about to find the trace of the Solen. This trace consists of nothing more than two small holes close together, sometimes broken into one, and presenting very much the appearance of the key-hole of a writing-desk. An experienced eye detects the trace with an unerring sagacity marvellous to the stranger, for the sand is perforated by holes of all sizes. The amateur may, however, ascertain which are the Solen-holes by attempting every one resembling a key-hole, and, after a few trials, he will gradually learn to detect them. If he have an iron rod, about as thick as a steel-pen-holder, let him place the point in the hole, and if the hole be the retreat of a Solen, the rod will pass some distance by its own weight. If any force be required to push it through the sand, he may be tolerably sure that he is at a wrong hole. It is in this way that fishermen habitually catch the Solen; and the amateur will find that considerable dexterity is required to use the rod with effect. It must be suffered to drop by its own weight till the fish be felt; then, by a half turn of the rod, the harpoon end fixes the shell, and the animal may be drawn up. Having repeatedly tried, and ignobly failed, I could not help admiring the dexterity with which my companion whipped them up, one after the other, scarcely ever missing; nor would my *amour propre* suffer me to quit the sands, until I had acquired sufficient skill to bring up a fish in about every three trials. This is the legitimate mode. It is the only one I find recorded in books; and from what Professor E. Forbes says, I conclude it is the only one known to naturalists.

But any one who hunts these Solens for sport, and

is less greedy of time than of amusement, will say that this mode sinks into insignificance beside the Jersey plan of "salting their tails." Having found a hole, we know that the Solen is at some distance underneath; it may be only a few inches, it may be many feet. The least disturbance will drive him irretrievably away. We must, therefore, allure him. Placing a pinch of salt over the hole, we await the result. In a minute or two the water begins to well up; this is succeeded by a commotion—the sand is upheaving—we hold our breath, and keep the hand ready to make a swift clutch —a final upheaval has taken place, and the Solen slowly shows the tip of his siphon; but he is still buried in the sand, and we must wait till he has thrust himself at least an inch above ground, or we shall lose him. It may be that, having come thus far, he will suddenly change his mind, and, instead of advancing, make a precipitate retreat. But if he raise himself an inch out of his hole, and you are swift, he is your prize. Sometimes, when clutched, he clings so firmly to the sand, that you break the shell or pull it out, and see half the torn body remain behind. At other times he will not appear at all. You have salted his hole, and after witnessing the preliminary commotions, you are nevertheless balked, for he retreats deeper and deeper, and his hole falls in. It thus appears that salting his tail does not necessarily imply a capture; and it is this uncertainty which gives a relish to the sport. Often when he has appeared at the hole, it is merely to see what is the matter, and to indulge in a not altogether frivolous curiosity as to the being who can illogically offer salt to *him* who lives in salt water; and he likes your appearance so little, that one glance is enough—he is off again like a shot. Two of them comported themselves in a very singular manner. They came to the opening, and defiantly throwing their heads at our feet, retired again in haste. Did they imagine we should be satisfied with *such* an offering? I picked up their heads, and moralised.

There is something irresistibly ludicrous in grave men stooping over a hole—their coat-tails pendant in the water, their breath suspended, one hand holding salt, the other alert to clutch the victim—watching the perturbations of the sand, like hungry cats beside the holes of mice; and there is something very absurd in the aspect of the queer Solen, poking up his inquisitive person; though *why* he is thus lured by the salt, I cannot guess. That he does not like the salt, is pretty certain, from his spontaneous decapitation under the infliction;* but why this should lure him is not intelligible. In conclusion, let me notice a passage in Mr Woodward's book, which not only contains an error, but implies that the salting mode of capture is not known even to well-informed naturalists. "Professor E. Forbes," he says, "has immortalised the sagacity of the razor-fish, who submits to be salted in his hole, rather than expose

---

* Strictly speaking, the Solen has no head at all. What is called the head, in the text, is simply the siphonal tubes, which are formed of muscular rings, placed longitudinally. In dissecting the Solen I found these rings spontaneously separate themselves in the water.

himself to be caught, after finding the enemy is lying in wait for him." * I suppose the " sagacity" was immortalised by Forbes in one of his playful moods; because not only is the fact on which the inference rests inaccurate, the Solen readily coming to his captor; but the Solen can have only slender pretensions to mental vigour of any kind.

Indeed, we are incessantly at fault in our tendency to anthropomorphise, a tendency which causes us to interpret the actions of animals according to the analogies of human nature. Wherever we see motion which seems to issue from some internal impulse, and not from an obvious external cause, we cannot help attributing it to "the will." No one seeing a bird snap at a fly with its beak, could doubt that the movement was voluntary; but if the bird's head were cut off, and the beak continued to snap, would not this throw a serious doubt on the voluntary nature of the former action? Yet this is what occurs with the curious *avicularium*, or "bird's-head process" of the Corkscrew Coralline: an animal doubtless familiar to many readers, some of whom have mistaken it for a Polype, it being indistinguishable from a Polype by the naked eye (Plate VII., fig. 4), although the microscope, revealing its internal structure, shows it to be a Polyzoon. The stem is twisted into a corkscrew shape, sufficiently remarkable to attract attention in rock-pools, or in tanks. On examining it attentively, it is generally seen to be furnished with a number of processes resembling vulture-

* *Manual of the Mollusca*, p. 15.

heads—one beneath each cup—having two mandibles, one fixed, the other movable by means of two sets of muscular fibres, visible within the head; and these mandibles keep up an incessant snapping, which occasionally entraps some worm, or minute crustacea, in an inexorable grasp. Very interesting it is to watch these bird's-heads snapping with vague vigour, while *above* them the animals to which they can scarcely be said to belong, are protruding from their cups; for, be it noted, the bird's-head does not form part of the animal, but issues from the stem on which the colony of animals abides; as if a gentleman residing in the parlour kept a watch-dog chained to his area gate. The position of this " process" has naturally led to the question, Is it an organ, or a parasite? The invariability of the position, and there being never more than one bird's-head to each animal, seem to point to its being an organ; but if so, what can be its function? Mr Gosse has suggested an ingenious answer : " Several observers have noticed the seizure of small roving animals by these pincer-like beaks; and hence the conclusion is pretty general that they are in some way connected with the procuring of food. But it seems to have been forgotten, not only that these organs have no power of passing the prey thus seized to the mouth, but also that this latter is situated at the bottom of a funnel of ciliated tentacles, and is calculated to receive only such minute prey as is drawn within the ciliary vortex. I venture to suggest a new explanation. The seizure of a passing animal, and the holding it in the tenacious grasp until it dies, may be the means of attracting the proper prey to the vicinity of the mouth. The presence of decomposing animal matter in water invariably attracts crowds of infusory animalcules, which then breed with amazing rapidity, so as to form a cloud of living atoms around the decaying body quite visible in the aggregate to the unassisted eye ; and these remain in the vicinity, playing round and round until the organic matter is quite consumed."*

The animalcules thus attracted would be whirled into the animal's mouth by means of its ciliated tentacles, and thus the bird's-head would be ancillary to the capture of food. Mr Gosse's explanation will equally hold good if the bird's-head be a parasite. One is naturally reminded of the analogous *Pedicellaria* of the Starfish.

Whatever may be the conclusion respecting these bird's-heads, the action of the bird being thus closely imitated puts us on our guard against the tendency to attribute psychological motives to the actions of animals. Indeed, unless we have previously assured ourselves of identity, or at least great similarity of structure, we shall always be in error when concluding an identity of function. Thus in the last chapter we saw how improbable is the supposition that the lower animals feel pain, in spite of our admission that all animals possess Sensibility. And everywhere the study of Comparative Anatomy teaches us, that before we can truly understand the Physiology and Psychology of animals, we

* GOSSE: *Tenby, a Sea-side Holiday*, p. 52.

must acquire the most minute knowledge of their structure. Much has already been done in this direction, but much more remains to do. There is still work for thousands of laborious students; and all genuine work will aid the science in its progress. Only let men observe with patient zeal, and forego the temptation to say they have seen what they have not seen, but what, from the report of others, they *expected* to see (a temptation which leads to the continuance of error more than any other cause),—only let the same veracity guide them with respect both to what is old and to what is new, equally deterring them from lightly supporting a current assertion, and from lightly promulgating a novel assertion,—only let this be done, and the humblest workers will bring their quota to the general fund. How much remains to be done may be gathered from the fact that, in the nervous system, which has been studied more than any other, recent investigations have led me to some unexpected results which, unless I greatly deceive myself, must profoundly modify current theory, and must assuredly modify our anatomical statements.*

We laugh at the German who is said " to construct the Idea of the Camel out of the depths of his moral

---

* The facility with which theories are extemporised by many who have little or no knowledge of the nervous structure, is only surpassed by the facility and confidence with which men attribute phenomena to electricity. It may be well, therefore, to state that our knowledge of the nervous system is at present in its infancy; we have not even established a secure basis; we have not established the primary *data*.

consciousness;" yet we are often really little better when we persist in *constructing* the forms of the simpler animals according to the analogies of our own structure, instead of directly *observing* what Nature presents to us. No sooner do we perceive certain manifestations of Sensibility, than we at once conclude the presence of a nervous system; no sooner have we a glimpse of a nervous system, than we are apt to conclude it to be the same in structure as our own. In like manner, Contractility is supposed to imply the presence of a muscular system; and the muscles of the lowest animals are supposed to be only minuter forms of the same organs found in man. But direct study of Nature assures us that there may be Sensibility without nerves, and Contractility without muscles; it further assures us that the nervous and muscular structures, when present, differ greatly in different forms of the animal series; and that we must understand these differences if we would understand Physiology. The law of unity of composition, when rightly understood, is a guide of great value; but it is too often a will-o'-wisp, luminously misleading. We get into the habit of naming organs according to a general idea of their functions, instead of keeping the idea of special function dependent on

---

To quote the emphatic language of one who has given his life to this subject: "Our knowledge even of the coarser framework of the nervous system is still too much in its infancy to permit us to venture, with any success, on the construction of theories respecting the functions of its various elements."—STILLING, *Ueber den Bau der Nerven-primitivfaser u. d. Nervenzelle*, 1856, p. 7.

special organ constantly before us. What would be said of a man who called his gig a gondola, his steamer a railway, his carriage a wheelbarrow? The *general* function of all these things is the same, but the *special* office of each differs from that of the others, and its structure bears reference to such specialty. In like manner a "proboscis" is not a "hand," although both are prehensile organs; nor are contractile fibres "muscles," although both subserve the same general function.

Descending from these abstract considerations to the particular subject now before us, namely, the nervous system, we note first, that minute and laborious as the researches have been, they have seldom taken the direction of comparative histology. The disposition of nerves and ganglia, and the structure of particular parts, such as the brain and spinal chord, have been studied with splendid results; but, as far as my reading extends, no one has thought of making minute and extensive comparisons of the various specialties of nervous tissue; and the reason has been that men have assumed a nerve to be always of one structure, forgetting Molière's humorous wisdom: *il y a fagots et fagots.*

When I first examined the nervous system of the *Doris* and *Pleurobranchus*, I was surprised to find the brain (œsophageal ganglia) of an orange-red, and yellow colour, instead of white or grey. The fact is familiar to anatomists, but the explanation given is more than questionable. Von Siebold attributes it to pigment scattered through the investing sheath (neuri-

lemma) and crowded on the surface of the ganglion; *
and Owen assimilates this pigment to "an arachnoid
membrane between the dense outer membrane and
the ganglions."† These explanations are easily disproved. On opening one of the ganglia (in a fresh
specimen) and pressing out the contents, I found that
the colour was not due to pigment in the membrane,
but to the contents of the ganglion, both cellular and
liquid; and by careful pressure, the whole contents
were ejected, leaving the colourless membrane behind.‡
Unimportant as this observation was, it was the starting-point of a long series of investigations. Finding
the contents of the ganglion were coloured, I inferred
that the coloured spots, irregularly distributed over the
upper portion of the nerve-trunks, and throughout the
bands connecting the ganglia into a collar, were also
due to ganglionic cells; this being proved, it followed
that the cells and granules of the ganglion were not
anatomically separable from the cells and granules of

---

* VON SIEBOLD, *Comparative Anatomy*, p. 233. So also DELLE
CHIAJE, *Istituzioni di Anatomia et Fisiologia Comparata*, p. 147:
"Nel centro principalmente è rosso-rancio, che ravvisasi pure nei
gangli. E circondato da valida membrana contenente molte glandulette giallastre."

† OWEN: *Lect. on Comp. Anat. of Invertebrata*, p. 550.

‡ The error is probably owing to a generalisation from the fact that
in many animals the pigment *is* distributed over the membrane.
LEYDIG (*Histologie d. Mensch. u. d. Thiere*, 1856, p. 50) confirms what
I have said in the text: "Diese Pigmentirung ist diffuser Art, sie
rührt her von einer rothen Flüssigkeit, welche das ganze Ganglion
durchtränkt, und nachdem des Neurilem eingerissen ist, in Tropfen
herausquillt."

the nerves; instead of the ganglion being a distinct structure from that of the nerve, the two were identical, the only difference being that the cells, both large and small, which predominated in the ganglion, became scarcer as the nerve was prolonged, and at last only made their appearance by ones and twos, amid the granular mass.

It may readily be imagined that the doubt thus raised, left me no peace. I dreamt of ganglia and their prolongations. They visited me in sleep, in half sleep, in noontide walking, and in evening reveries. For weeks I was in a "nervous fever." Every animal I could lay hands on was sacrificed to the inexorable scalpel,—from the new-born Puppy, to the Bee which flew into my study; from the Lizard, Frog, and Toad, to the Grasshopper and Locust; from the hideous Dogfish, to the graceful Pipefish; from the Hermit crab, to the Dragon-fly larva; from the garden Snail, to the Slug; from the Solen, to the Sea-hare. Scores of dissections only increased my doubts respecting current doctrines; since, whether fresh specimens, or preparations after treatment with alcohol, with chromic acid, and with acetic acid, were under examination, the facts required by the doctrines were not discoverable, but, in lieu thereof, facts altogether opposed to them.

The physiologist will read with surprise of the absence of fibres in the nerves, which are universally held to be simply fasciculi of fibres; and, indeed, the discovery so much surprised me at first, that it was long before I could persuade myself it was no optical illusion, or the result of disintegration in the nerve itself; but having examined both fresh and prepared specimens, in great quantities, I affirm, that in the genera *Doris*, *Pleurobranchus*, *Aplysia*, *Solen*, and *Limax*, the nerves are for the most part totally destitute of fibres.*

Within the investing sheath of areolar tissue † is contained a mass of granules, cell nuclei, and occasional cells large and small, but not one primitive nerve-fibre. If we compare the structure of the nerves in the Doris with that in the garden Snail, we shall immediately perceive the difference, the latter animal having distinctly fibrillated nerves, the former nothing but amorphous nerve-substance. At first it occurred

---

* Dr Inman, of Liverpool, informs me, while these sheets are passing through the press, that he has examined the nervous system of three very fine specimens of *Dendronotus arborescens* in which he finds precisely the disposition I have described: "The nervous masses were made up of very large cells, $\frac{1}{15}$ inch in diameter, full of granular matter, and surrounded by a thick coat of granular material of a brownish colour. These were most numerous in the four large ganglia, but were found sparingly in the main nervous trunks. Some cells or vesicles were much smaller than others, the smallest about $\frac{1}{45}$ of an inch. The nervous trunks were filled with granular matter, in which no appearance of fibrillation could be traced, but on squeezing out the neurine it was found that there was an abundant supply of vesicles of various sizes, as well as of granular matter. In one main tube at some distance from the ganglia there was a faint appearance of fibres which were very small in length and breadth, and quite solid." This last fact is very interesting, and agrees with what I have noted further on of the dragon-fly larvæ.

† It may not be unnecessary to warn the amateur who may verify these observations, not to confound the fibres of the investing sheath with nerve-fibres.

to me that the granular structure might be peculiar to the molluscs without shells; but the bivalve Solen and the Aplysia contradict such a suggestion.

Again, in the brains of several young Tritons, or Water-Newts, I could find no fibres; none in that of a young Frog; few in that of an adult Toad; none in that of adult Pipefish (*Syngnathus*); none in that of a Dogfish (*Acanthias*) of about a foot and a half long. It would be perilous to assert that there were absolutely *no* fibres in the brains of these animals, but if there were any, they must have been exceedingly rare, as I could not find one: and theory requires that there should be one or more for every cell.*

These facts are important as well as novel, and force us to the conclusion that fibres are not *necessary* to the conduction of nerve-force, although they may be *special* organs of conduction, wherever they exist. The point to which the reader's attention is required is that nerve-force can be transmitted—that nerve-actions can take place—in the absence of primitive fibres. This might have been concluded from the structure of the olfactory nerve alone, which, by a remarkable peculiarity, shows a break in the continuity of its fibres, the intervening space being occupied with granules only. This is the case in all animals.† And I find a still

---

* Plate VII., fig. 2, gives a hypothetical diagram of the relation of fibres and cells in the spinal chord and brain, according to the latest ideas. Whatever truth it may have in reference to the higher vertebrates, it is certainly not applicable to those fish and reptiles I have examined.

† See the description of the olfactorius in LEYDIG, *op. cit.*

greater break in the sympathetic nerve of the new-born puppy. As the trunk is about to join the ganglion, the fibres disappear, and give place to granules; nor do fibres appear in the ganglion at all.* Now, as theory requires every nervous impression to be conveyed by a fibre *to* a cell and by a fibre *from* the cell, we see at once that the foregoing facts, or any one of them, must strike at the root of such a theory, nervous impressions being indubitably transmitted where no fibres exist, and where a solution of continuity exists.

Since these observations were first published, I have found more than one professional man disposed to doubt that the theory of conduction is held by authoritative physiologists; it will be useful, therefore, to show that I was not combating an exploded error. In England Dr Todd certainly holds place in the highest ranks; and in 1856 he thus wrote of softened brains: "In all cases the cerebral disease reaches such an extent that the *vesicular matter imperfectly generates the nervous force*, and the *fibrous matter* becomes a *bad conductor of it*, or even a non-conductor, or *its continuity is interrupted, and so its power of conduction is rendered,*

---

* The necessity for caution, both in extending our observations and in making them public, is illustrated by the fact, that since the above was written, I have seen *one* sympathetic ganglion in which the fibres *did* penetrate; and it is worth mentioning that I found the continuity of the fibres uninterrupted in all the sympathetic nerves of a new-born kitten. Let me add that, except in the case of the Slug (*Limax*), all the facts stated in the text are founded on multiplied observations; the magnifying powers used being 350—750 linear, with one of Smith and Beck's excellent Microscopes.

in the one case, physiologically, in the other mechanically, impossible." Can anything be more explicit? A solution of continuity between fibre and cell is said to render the conduction of nerve-force impossible; yet I have shown that such a solution is constant in many animals, and is present in the embryonic condition of all animals. No one even slightly acquainted with the present state of science will fail to see the bearing of this fact.

But before we proceed further it will be requisite to ascertain, as far as possible, whether the facts I have discovered had been already made public. When Europe furnishes its hundreds of diligent workers in any department, no one can expect to stand in isolated originality; he must be prepared to find that others have more or less anticipated him. I had no sort of doubt that the facts, which to me were full of significance, must have been observed by others; but I was persuaded that no one had seen their significance, because no one expressed a doubt respecting the theory which they undermine.† My first step was to send to England for Leydig's work on Comparative Histology, the latest authoritative publication. There, indeed, I

---

* TODD: *Clinical Lectures on Paralysis*, 2d ed., 1856, p. 211. Equally explicit is GRATIOLET: *Anat. Comp. du Système Nerveux*, 1857; ii. p. 4. So also KIRKES and PAGET, in their *Handbook of Physiology*, p. 364. "All nerve-fibres are mere *conductors* of impressions." In SEGOND, *Traité d'Anatomie Générale*, p. 111, the nerve-fibre is said to manifest a new and characteristic property, that of *transmissibility*, as the muscle-fibre manifests that of contractibility. Compare also FUNKE, *Lehrbuch der Physiologie*.

† We need only turn to FUNKE's *Lehrbuch der Physiologie*, one of the ablest and most erudite, as well as the latest of treatises, to be assured of this.

found the existence of granular nerves stated as a fact, though without specific information either respecting their discoverer or the animals in which they existed; and without a hint of any physiological significance in the fact.* On my return home I made diligent search, and by means of Canstatt's *Jahresberichte* for 1854 (p. 66), learned that Meissner had discovered granular trunks in the thread-like tiny worm *Mermis;* and H. Müller and Gegenbaur in the naked gasteropod *Phyllirhoë.*

On procuring the memoirs referred to,† I found the fullest confirmation of my own observations, but no appreciation of their physiological significance. Müller and Gegenbaur say: "Distinct fibres are not discoverable in the trunks, which appear to consist of nothing but a clear granular streaky substance (*aus einer hellen feinkörnig streifigen Substanz*). In some instances there were small groups of ganglionic cells." And this is *all* they remark.

Meissner's observations are given in greater detail, and appear to have suggested doubts, as analogous observations did to me. "The four trunks," he says, "which issue from the ganglia have at first a clearly fibrous structure, so that at the torn ends single fibrillæ appear; but these fibres in their course soon melt into a homogeneous band in which no trace of fibre remains." Curiously enough, the branches given

---

* LEYDIG, *Histologie*, pp. 59, 185. "Die Nervensubstanz ist entweder mehr homogen und molekular, oder mehr von faserigem Aussehen."

† See SIEBOLD u. KÜLLIKER'S *Zeitschrift f. Wissen. Zool.*, v. p. 233, 360; and vii. 99.

off from these trunks, although they commence as homogeneous bands, presently break up into fibres which continue to the peripheral termination. In presence of such a structure, Meissner could scarcely have missed the suggestion it forced upon him: "In *Mermis* the anatomical proof is easy that a conduction from the periphery to the centre must take place by some other means than that of a completely isolated and throughout equally constructed fibre." * On reading this sentence, I fancied that the same idea must have occurred to him as to me, and that he would follow it up by some further observations, or at any rate by some physiological reflections. Two years later, however, he published another Memoir,† wherein he notices this peculiarity of the nerve-trunks as "certainly very important for the physiology of the nervous system;" but instead of seizing its true significance, he proposes an explanation which would never have been proposed if the facts I have observed had been known to him; for, confining himself to the peculiar structure of the trunks and branches of these worms, he suggests that "the trunks must not be regarded as trunks, as anatomical bundles of isolated fibres, but as *peculiar and intermediate conducting organs interposed between the central organ and the peripheral nerves.* If they are really to be regarded as simple nerve-trunks, in the sense in which the word is used as respects higher animals, we ought to find them composed of fibres—which is not the case." I need scarcely criticise such a sug-

* *Loc. cit.*   † *Loc. cit.* vii. 99.

gestion; the mere fact of a sympathetic ganglion being connected with a nerve which, although fibrous, has its interspace of granules, is enough to destroy the hypothesis; not to mention the fact of so many animals being without fibres at all.*

From what has been already said, the conclusion is inevitable that the *conduction of nerve force does not take place by means of fibres only.* The fibres *may* be special organs of conduction, and as special organs, a corresponding specialty of function must be assigned to them; and into this we must now inquire.

Let us assume that the homogeneous nerve transmits the impression in a mass, just as the sounding-board of a piano, if struck, will yield a certain resonance; but the fibrous nerve will transmit the impression along each separate fibre, like the sounding-board when struck by keys; the amount of nervous impression and the amount of sound in each case may be equal, but the varieties and combinations possible to the latter are impossible to the former. Or, to vary the illustration, let us assume two men to be equally susceptible to the general effect of colour, but one of them, an artist, to have more susceptibility to the minute differ-

---

* Meissner's observations furnish a very noticeable fact, namely, that while in *Mermis Albicans* the trunks are homogeneous, in another species, *Mermis nigrescens*, they are fibrous! In the face of such evidence no single exception to the facts I have stated would surprise me—for instance, that fibres could be found in a *Doris*, a *Pleurobranchus*, an *Aplysia*, or a *Solen*—such exceptions would in nowise invalidate my conclusions, for which, indeed, one single case of non-fibrillated nerve would be ample evidence. [I have since found fibres in the nerves of a Doris.]

ences of colour; although the nervous impression may be equal in the two, it will be less homogeneous in the artist, whom we may suppose to have a more specialised retina.

The assumption that fibres are organs of conduction at all, may be disputed; nor, if what was previously said respecting the identity of cell and fibre, in ultimate structure, and of the identity of ganglion and tube, be admitted, can we allow the old hypothesis of conduction to be more than a metaphor. The notion of an actual conduction taking place, analogous to the conduction of electricity, is extremely doubtful to me. If the nerves are identical in elementary structure with the ganglia, and consequently *participate* in the properties of the ganglia, they can no longer be regarded as the conducting-rods of the battery, but as essential parts of it. In our present ignorance of the true process we may continue to employ the metaphor of conduction, if we understand by it simply the change which follows when a nerve is affected; and we may then gain some glimmering of the special function of the fibres, and the meaning of their increase with old age. Nerve-tissue in its earliest stage is wholly without fibres; as development advances, the fibres multiply.* In old age the brain hardens from excess of fibres, as the bones harden from excess of lime; so that what

* Not till the beginning of the fourth month of the human embryo are fibres discoverable in the spinal chord.—TIEDEMANN, *Anatomie du Cerveau*, p. 126. When the fibres first make their appearance in the brain, I know not, but in the brains of a new-born puppy and kitten I could find no trace of them. Indeed, the naked eye showed that no

originally constituted a source of strength becomes a source of weakness. Probably to this predominance of fibres may be assigned the incapacity of acquiring new ideas in old age. Intellectual vigour is often manifested by men of a very advanced age, but the vigour is shown in dealing with old trains of thought, not in originating new. To assume a new attitude of thought, it may be necessary to develop new fibres; and this cannot be done in a tissue already too fibrous. A similar hypothetical explanation suggests itself for the formation of fixed ideas, monomanias, habits, and tendencies.

But I will not venture further into this hypothetical region, the few anatomical facts hitherto ascertained presenting too narrow a basis for such speculations. One embryological indication may, however, be added. The nerves of insects are, it is known, distinctly fibrous (although in the bee and locust I have observed the fibres occasionally melting into mere granules), but in the larvæ of insects the nerves are often mostly granular. Thus in the active predatory Dragonfly Larva—the water-tiger, as it is called—I found the great ventral chord possessing distinct fibres, but in many places it was purely granular, the granules not having even a linear disposition.* In the preparation I have made of this object a very interesting analogy between the development of nerve and muscle is presented. Muscles

differentiation had as yet taken place between the grey and white matter; and the Microscope, with a power of 750 linear, confirmed this impression, the structure of both grey and white matter being wholly vesicular and granular.

* In more advanced Larvæ these chords are wholly fibrous.

are of two kinds, striped and unstriped, the former being generally, but erroneously, called voluntary, the latter involuntary muscles. According to recent researches, it has become evident that the striped muscle is only a more differentiated form of the unstriped, there being several intermediate stages between the two.* In the preparation I have made of the ventral chord of the dragonfly larva, this is strikingly exhibited. A fragment of muscle is attached, the fibrillæ of which, instead of being striped (all the muscles of insects are of the striped kind), are partly striped, partly unstriped; that is to say, in the same bundle some of the fibrillæ are without the transverse markings, and those fibrillæ which have such markings have them only part of the way down, the remainder of the fibrillæ being unstriped. This is not only interesting as a fact in muscle development, but presents a striking analogy to the development of nerve-fibres, which we here see in the same trunk partly emerged from their primitive granular condition.

I conclude, therefore, that the differentiation of nerves shows the following phases: 1st, As in many Molluscs and all embryos, a granular homogeneous mass; 2d, As in insects, and perhaps Crustaceans, a linear disposition of the granules into fibres, but without an investing sheath; 3d, Fibres, or rather *tubules*, differing from the preceding in structure, having each an enveloping sheath, which isolates one fibre from the other, so that the nerve becomes a fasciculus of tubules.†

---

* LEYDIG, *Histologie*, p. 43.

† STILLING, *op. cit.* p. 11-13, decides that *all* primitive fibres have an investing sheath; but unless he would deny the claim of those named

It would lead us too far to follow the many applications of these facts to the vexed questions of nervous histology and physiology. The hotly-debated controversy respecting fibres as prolongations from ganglionic cells, for example, seems to me decisively settled by the fact that, in the Molluscs, we have cells without fibres, and by the fact that, in the recently-born dog, we have fibres which have not yet effected a junction with the cells. Again, when Funke, reviewing the controversy respecting the existence of ganglionic cells destitute of processes, says that, " from all we know of the functions of the nerve-elements and the laws of conduction in them, an isolated apolar nerve-cell appears as an anomaly (*Unding*) to which we can in nowise assign a physiological purpose,"* he is assuming that without fibres terminating in cells no nervous transmission can take place—an assumption flatly contradicted by the facts we have just been considering.

What we metaphorically call " nervous conduction " takes place not only in the absence of fibres, but also in the absence of *any nerves whatever*. There is nothing like the sharp angle of a paradox to prick the reader's attention ; and here is one in all seriousness

in the second class to be considered as fibres, he is certainly wrong. It should be noted that what are called fibres in the vertebrate nerve-tissue are really tubules with fluid contents. This fact was discovered by LEEUWENHOEK, who describes them as vessels (see his *Select Works*, ii. 303), a discovery often erroneously attributed to EHRENBERG, to whom the credit is given in Mr HOGG's popular work, *The Microscope*, 3d edition, p 525.

* FUNKE, *Physiologie*, p. 419.

presented to him. The fact is demonstrable, that both Contractility and Sensibility are manifested by animals totally destitute of either muscles or nerves. Some physiologists, indeed, misled by the *a priori* tendency to "construct" the organism in lieu of observing it, speak of the muscles and nerves of the simplest animals; because, when they see the phenomena of Contractility and Sensibility, they are unable to dispossess themselves of the idea that these must be due to muscles and nerves. Thus, when the fresh-water Polype is seen capturing, struggling with, and finally swallowing a worm, yet refusing to swallow a bit of thread, we cannot deny that it manifests both Sensibility and Contractility, unless we deny these properties to all other animals. Nevertheless, the highest powers of the best microscope fail to detect the slightest trace of either muscle or nerve in the Polype. Leydig, indeed, describes and figures what he calls the muscles of the Hydra.[*] But what are they? A network of contractile cells of irregular shape, such as Ecker [†] had previously described as the "unformed contractile substance," with this difference, that Leydig discovers a nucleus attached to the wall of each cell. I have seen it often, but the observation by no means warrants the conclusion that this proof of the network being cellular, is a proof of its being muscular. The cells resemble muscles in no respect, except that

---

[*] *Op. cit.*
[†] ECKER, *Zur Lehre vom Bau und Leben der contractilen. Substanz der niedersten Thiere.*

they are contractile; and we are not warranted in calling every contractile cell a muscle-cell; otherwise we must call the unicellular animals muscles.* The Hydra presents us with contractile substance (or cells), but to call the contractile substance a "muscle," is to outrage language more than if a wheelbarrow were spoken of as a railway locomotive; and even this latitude of language will not serve our turn with respect to the nervous system of the Polype, since nothing resembling a nerve or nerve-substance is discernible in it. We must either deny that the Polype manifests Sensibility, or we must admit that Sensibility may exist without nerves.

In presence of these facts, physiologists, who cannot conceive Sensibility without a nervous system, but are forced to confess that such a system is undiscoverable, assume that it exists "in a diffused state." I have noticed this illogical position in a former chapter. It is a flat contradiction in terms: a diffused nerve is tantamount to a liquid crystal; the nerve being as specific in its structure, and in the properties belonging to that structure, as a crystal is. Now, this specific structure—or anything approaching it—is not to be found in the Polype.†

* VICTOR CARUS and REICHERT have discovered contractile cells in the vitellus, but no one would call these muscles.

† "Sarebbe una vera perdita di tempo," says DELLE CHIAJE, "per colui che volesse ricercare nervi negli animali Infusori, nei Polipi, nelle Meduse e nelle Actinie."—*Istituzioni di Anat. e Fisiolog. Comparata*, i. 118. He denies the existence of nerves even in the Holothuriæ. So likewise does VOGT, *Zoologische Briefe*, with several other anatomists.

Whence, then, is the Sensibility derived? Either we must admit the presence of what cannot be discovered; or we must admit that a function can act without its organ; or, finally, we must modify our conception of the relation between Sensibility and the Nervous system. Which of these three conclusions shall we adopt?

Not the first: for, to admit the presence of an organ which cannot be discovered, even by the very highest powers, although easily discoverable in other animals by quite medium powers, would be permissible only as the last resource of hypothesis, when no other supposition could be tenable.

Not the second: for philosophic Biology rejects the idea of a function being independent of its organ, since a function is the activity of an organ. The organ is the agent, the function the act—a point to which we will presently recur.

The third conclusion, therefore, seems inevitable: we must modify our views. But how? Instead of saying, "Sensibility is a property of nervous tissue," we must say, "*Sensibility is a general property of the vital organism which becomes specialised in the nervous tissue in proportion as the organism itself becomes specialised.*"

We have no difficulty in understanding how Contractility, at first the property of the whole of the simple organism, becomes *specialised* in muscular tissue. We have no difficulty in understanding how Respiration, at first effected by the whole surface of the simple organism, becomes specialised in a particular part of that surface (gills or lungs) in the more complex organisms; nor should we have more difficulty in understanding how Sensibility, from being common to the whole organism, is *handed over to a special structure,* which then performs that function exclusively, as the lungs perform that of Respiration, or the muscles that of Contraction. Nay, more: just as animals possessing special organs for Respiration, do also, in a minor degree, respire by the general surface, so, according to my observations, it is almost demonstrable that animals possessing a special nervous system also manifest Sensibility in parts far removed from any nervous filament. In the higher animals this is probably not the case.* The division of labour is more complete. The stomach digests, the glands secrete, the muscles contract, and the nerves feel. Of course, the power is greatly increased by this division of labour; the more complex the organism, the more various and effective each function. In the pregnant language of our most thoughtful poet,—

> "All Nature widens upward. Evermore
> The simpler essence lower lies;
> More complex is more perfect, owning more
> Discourse, more widely wise." †

---

* I say *probably*, because recent investigations have shown that parts which, in the normal healthy condition, are absolutely insensible, such as tendons, ligaments, the dura mater, and the periosteum, become intensely sensitive in a state of inflammation, and this cannot be attributed to the nerves.—See FLOURENS, in *Annales des Sciences Naturelles*, 1856, IV. Série, vi. 282; and compare Dr INMAN'S work on *Spinal Irritation*. Further, Mr TOMES has communicated to the Royal Society a paper on the "Soft Fibrils in the Dentinal Tubes," which shows a sensitive structure, *not* nervous, in the teeth.

† TENNYSON: *The Palace of Art.*

It is truly remarkable that the zoologist who claims the merit of having originated this conception of the "division of labour" as a law in the organic economy,* should be among the stanchest defenders of the old metaphysical idea that functions are not dependent on organs; and as this question is not only important in itself, but of interest in the present discussion, it may detain us for a moment. The argument, as conducted by Milne Edwards,† is irresistible, because in it he confines himself to showing that special organs may disappear, and the general function nevertheless remain; for instance, that lungs and gills may be absent, but the function of Respiration will still be present: "C'est une erreur grave de croire," he says, "qu'une *faculté determinée* ne puisse s'exercer qu'à l'aide d'un seul et même organe." The grave error appears to me wholly on the side of those who hold the contrary opinion. The reader will perceive that when Milne Edwards concludes, "que la fonction ne disparait pas lorsque l'instrument spécial cesse d'exister," the eminent zoologist is guilty of a logical mistake very frequent in biological discussions—the mistake of confounding the *general* with the *particular*. Thus an

---

* MILNE EDWARDS: See his *Introduction à la Zoologie Générale*. The conception, however, belongs to GOETHE, *Zur Morphologie*, 1807; *Werke*, xxxvi. 7—the French naturalist having the merit of application and abundant illustration of the law.

† *Loc. cit.* p. 60, and *Leçons sur la Physiologie et l'Anatomie Comparée*, i. 22. "Les faits dont je viens vous entretenir montrent combien sont fausses les opinions de quelques naturalistes qui admettent comme une sorte d'axiome physiologique que la fonction dépend toujours de son organe."

animal may possess the general function of Locomotion, without possessing the special function of walking or flying; it may have the general function of Sensibility, without the special Senses; the general function of Assimilation, without special Digestion; for the *special* functions special organs are required, legs or wings, eyes or ears, intestinal canal, and so forth.

The peculiar faculty of locomotion known as "flying," can only be performed by the peculiar organs known as "wings;" but the general faculty of locomotion can be performed by a simple contractile tissue. As soon as we disengage language of its ambiguity, the truth is easily seen: the appearance of each special organ in the animal series is coincident with the appearance of a special and corresponding function; or, descending the scale, with the disappearance of each special organ, the corresponding specialty of function disappears. In other words, Function is dependent on Organ, as the Act is on the Agent. Would it not seem wholly preposterous to say that railway locomotion was not necessarily dependent on railways, because, before railways were invented, goods and passengers were transmitted by waggons and coaches? Does not every one see that the special form of railway transmission gives it a power, velocity, and extent, wholly unattainable by waggons and coaches; and that this power, velocity, and extent, are due entirely to the peculiarity of the methods of transit? The railway differs from the tramway, and the tramway from the old coach-road, in special modifications, as the lungs of a mammal differ from the gills of a fish, and the gills from a merely respiratory surface in the Zoophyte. To say that the *same* function is performed by all, is to confound the general with the particular; and to say that functions are consequently independent of organs, is worse than this, because it leaves out of sight the fact that the whole surface of the animal which respires is the "organ" of that general respiration. When acts can be performed without agents, and animals can exist without bodies, in the shape of pure syntheses of functions, then will it be a "grave error" to conclude that functions are necessarily dependent on organs, and not till then.

Although we have been forced to admit that Sensibility can be manifested without nerves, and the paradox therefore of nervous conduction taking place without nerves was only a paradox in its terms, yet inasmuch as functions are necessarily dependent on organs, we are also forced to conclude that the various specialisations of nerve-tissue must bring with them corresponding specialisations of property. What those are we know not, perhaps may never know; but the mere recognition that such things must be, will help us to understand many points. It explains, for example, the absence of Pain in the lower animals, spoken of in the last chapter; it explains the possibility of such myriads of subtle differences in the perceptions of men so nearly allied as twin brothers.

And now, dear reader, we must part company, after

having wandered together over many a reef and bay, and skirted the great plains of Philosophy, traversed by many foot-sore pilgrims who beckon us to follow. We have caught some glimpses of the marvels and delights which await us in every rock-pool; and we have seen how a mere amusement will naturally lead us into the solemn temples of Philosophy.

The naturalist may be anything, everything. He may yield to the charm of simple observation; he may study the habits and habitats of animals, and moralise on their ways; he may use them as starting-points of laborious research; he may carry his newly-observed facts into the highest region of speculation; and whether roaming amid the lovely nooks of Nature in quest of varied specimens, or fleeting the quiet hour in observation of his pets—whether he make Natural History an amusement, or both amusement and serious work—it will always offer him exquisite delight. From the schoolboy to the philosopher, all grades find in it something admirably suited to their minds. It brings us into closer presence of the great mysteries of life; and while quickening our sense of the infinite marvels which surround the simplest object, teaches us many and pregnant lessons which may help us through our daily needs.

In suggestive but untranslatable verse, Goethe asks what higher aim can we have here on earth than to trace the revelation of the Divine in Nature; how Fact becomes etherealised into Thought, and how Thought in turn becomes incorporate in Fact—as

Greece, which was once a mere geographical boundary, containing a handful of men, has been ever since a spiritual influence moulding the destinies of the world; or, conversely, as the mere idea of a locomotive engine which arose in the mind of George Stephenson, passed into the gigantic fact of the Railway system.

> " Was kann der Mensch im Leben mehr gewinnen
> Als das sich Gott-Natur ihm offenbare,
> Wie sich das Feste lässt zu Geist verrinnen,
> Wie sich das Geisterzeugte fest bewahre." *

All the forms and facts of Nature carry with them a deep spiritual significance, and cannot be reverently studied without revealing it; for are they not the manifestations of the Universal Life? Unreflecting minds often deem it a trivial occupation for serious men to devote themselves with patience to the study of anatomical details, and the scrutiny of facts which seem to have no practical bearing on the great affairs of life. These details, like all other facts of Nature, may, indeed, be studied in a trivial spirit, uninspired by a loftier aim; but under their lowest aspect they have still the inalienable value attendant upon all truth; and under their highest aspect they teach us something of a noble wisdom which profoundly affects the practical affairs of life, by affecting the direction and the temper of our thoughts.

* GOETHE: *Bei Betrachtung von Schiller's Schädel.*

GLOSSARY.

ABRANCHIATE—Without branchiæ, or gills.

ACALEPHÆ—Jellyfish.

ACTÆON—A shell-less Mollusc, one of the "sea slugs."

ACTINIA—The Sea Anemone. So called from the ray-like disposition of its arms or tentacles.

AMŒBA—A microscopic animalcule, of jelly-like substance, without organs, without even constant form.

ANASTOMOSE—When the mouths of two vessels unite and blend together, forming one continuous vessel, they are said to anastomose.

ANNELID—Almost all worms are composed of a succession of rings or segments; hence they are included under the term *Annulata* or *Annelida*.

ANTHEA—A species of Sea Anemone.

APLYSIA—A gasteropodous Mollusc, popularly called the Sea-Hare (Plate II. fig. 3.)

ARBORESCENT—Branched like a tree.

ASCIDIAN—A Mollusc, without head or shell, having two orifices nearly on a level, and shaped like a leathern bottle (*askos* is the Greek for flask). See Frontispiece, fig. 4. The Ascidians are also found living in colonies. Hence there are Solitary Ascidians and Compound Ascidians.

AURORA—A species of Sea Anemone, orange-tentacled.

AUTOMATIC—Actions are said to be automatic when they occur without the intervention of the will—*e. g.* the heart acts automatically; we breathe automatically.

BALANI—The *Balanus* is one of the fixed *Cirripeds*. It is not a "shellfish"—*i. e.* a Mollusc—in spite of its resemblance to one. The rocks are often covered with their small flat and conical shells. They are also found on the shells of the Whelk and other molluscs.

BIOLOGY—The Science of Life, including Botany, Zoology, Physiology, and Anatomy, which are the branches of this general subject.

BIVALVE—Molluscs which have two shells closing together are called Bivalves—*e. g.* the Mussel or the Oyster.

BOTRYLLUS—A little animal belonging to the Compound Ascidians. Its colonies are found growing on sea-weed and stones, and look like masses of jelly, with stars imbedded in them.

BRANCHIÆ—The gills.

BRITTLE-STAR—One of the Starfishes.

BRYOZOA—See *Polyzoa*.

BYSSUS—Silky filaments by which Mussels and other shell-fish attach themselves firmly to the rocks.

CÆCAL—The *cæcum* is a blind tube; cæcal prolongations of the intestine are therefore ramifications without openings at the farther ends.

CALAMARY—One of the Cephalopoda (*Loligo*), vulgarly called "Squib," and often confounded with the Cuttle-fish.

CEPHALIC—Belonging to the head.

CEPHALOPODA—Molluscs having long prehensile arms projecting from the head—*e. g.* the Cuttle-fish.

CILIA—Microscopic hairlike bodies which constitute the sole organs of locomotion in some animals, and which, by their incessant vibration, cause currents in the water. Ciliary action has nothing voluntary in it, and continues long after the death of the animal.

CILIOBRACHIATE—Animals having cilia on their tentacles. They were formerly supposed to be Polypes, and distinguished only by the presence of cilia on the tentacles. They are now classed as *Polyzoa*.

CIRRIPEDS, or CIRRIPEDA—A class of articulated animals having curled and jointed feet.

CLAVELLINÆ—Compound Ascidians. (Frontispiece, fig. 4.)

CRASSICORNIS—A species of Sea Anemone, also called the *Coriaceous Anemone*.

CRUSTACEA—A class of articulated animals having a *crust*, or thick skin, which they shed periodically—*e. g.* Lobsters, Shrimps, Water-fleas.

DAISY—A species of Sea Anemone: *Actinia Bellis*.

DIANTHUS—A species of Sea Anemone: the Plumose Anemone.

DIFFERENTIATION—When one tissue—*e. g.* cartilage—is converted into another—*e. g.* bone—it is said to be differentiated. The process by which the fluid mass that constitutes an egg becomes converted into a bird, is the process of differentiation. *Growth* is mere increase of bulk; *differentiation* is the appearance of new matter or new forms.

DIŒCIOUS—Plants or animals of separate sexes.

DORIS—A Mollusc of the Nudibranch order. (See Plate II., fig. 2.)

ECHINODERMATA—Animals having spinous coverings, as Starfishes, Sea-eggs, Sea-urchins, &c.

ENDOSMOSIS—The passage of one fluid, through a membrane, to another fluid.

ENTOMOSTRACA—Little Crustaceans, popularly called Water-fleas.

ENTOZOA—Parasitic animals which live *inside* the bodies of other animals.

EPIZOA—Parasitic animals which live *outside* the bodies of others.

EOLIS—A Mollusc of the Nudibranch order. (See Plate II., fig. 1.)

EPITHELIUM—A delicate membrane covering mucous membranes, as the epidermis covers the skin.

GANGLION—A centre of nervous matter from which nerves radiate. The spinal chord is a series of ganglia blended together. So is the brain.

GASTEROPODA—Molluscs which crawl on their ventral surface—*e. g.* the Slug, Snail, &c.

GONIASTER—One of the Starfishes.

HEPATIC—Belonging to the liver.

HERMAPHRODITE—Having both sexes combined in a single individual.

HISTOLOGICAL—Histology is the doctrine of the tissues; and tissues are the webs out of which the organism is fabricated. Organs are composed of various tissues.

HOMOLOGUE—The same organ in different animals, having identity of parts with variety in form and function. The wing of a bird is the homologue of a man's arm, and of a whale's fin. But a gill is the *analogue* of a lung, not the homologue.

HYDRA—Fresh-water Polype.

LARVA—The insect before its transformation.

LARVIPAROUS—Producing offspring by Larvæ.

LERNÆA—An order of crustacean parasites.

LIMAX—The slug.

MEDUSÆ—Jelly-fish.

MOLLUSCS—A division of the animal kingdom embracing great varieties—*e. g.* the Oyster, Cuttle-fish, Snail, &c.

MORPHOLOGICAL—Morphology is the doctrine of the changes of form which organs undergo during development.

NAÏS—A worm very common in ponds; also found in salt water.

NEURILEMMA—The investing sheath of the nerve.

NUDIBRANCHS—An order of Gasteropod molluscs having their gills exposed—*e. g.* Doris, Tritonia, &c.

ŒSOPHAGUS—The tube leading from the mouth into the stomach.

OVARY—The organ in which the eggs are formed.

PARENCHYMA—The soft tissue of organs; generally applied to that of glands.

PARIETES—The walls of the different cavities of the body.

PARTHENOGENESIS—The production of offspring from unfertilised eggs—*i. e.* from virgin mothers.

PECTEN—A bivalve Mollusc—the *Scallop*.

PEDICELLARIA—A curious organ—perhaps a parasite—found growing round the spines on the skin of the Starfish. When magnified it looks like a pair of pincers.

PEDICELLINA—A Polyzoon.

PLANARIE—Flat worms without segments, mostly parasitic, abundant in ponds and on sea-weed.

PLEUROBRANCHUS—A gasteropodous mollusc, with its gill hanging down from the side.

POLYZOA—A class of animals long held to be Polypes, which externally most of them resemble; now ranged under the head of Molluscs. They are also called *Bryozoa*, especially by foreign writers. The *Plumatella* (Frontispiece, fig. 1) is a Polyzoon.

PYCNOGONIDÆ—Small crustaceans, mostly parasitic.

SABELLA—One of the worms which live in tubes; but its tube is not calcareous; it is formed of mud or particles of sand glued together.

SAGITTA—An animal not yet classed. (See p. 251, and Plate V., fig. 2.)

SALPA—A mollusc, of the order *Tunicata*, which floats in the open sea, and produces offspring chained as it were together.

SERPULE—Worms which live in calcareous tubes secreted from their own bodies. In the water they expand lovely feathery gills.

SPERMATOZOON—A microscopic filament, forming the essential part of the fertilising fluid of animals, and formerly supposed to be an animalcule. Experiments have proved that without spermatozoa no fertilisation takes place; but that spermatozoa separated from the fluid suffice.

SPICULA—Fine-pointed bodies like needles.

TEREBELLA—A worm living in a tube of sand or shells glued together by an exudation from its body. (Plate VII.)

TROGLODYTES—Cave-dwellers; a species of Sea Anemone.

VENUSTA—A species of Sea Anemone: orange-disked.

VITELLUS—The yolk of the egg.

INDEX.

ABRANCHIATE, the Eolis an, 112.
Acalephæ, entrance of water into the blood in, 117.
Acanthias, no fibres in the brain of the, 394.
Actæon, the, 216, 255—peculiarity of the ova of, 260 *note*.
Actinia, the finding of the first, 13, 14—vitality of separated filaments of the, 60—name of, first adopted, 124—longevity of, 146 *note*—anatomy, 164. *See also* Anemone.
Actiniæ, the, 13—jars suitable for, 35—entrance of water into the blood in, 116—how their animal nature is determined, 127 *et seq.*—viviparous, 148—reproduction among them, 163 *et seq.*—their organisation, 167—are they of separate sexes? 172—their so-called digestion or assimilation, 226 *et seq.*—species of, found at Scilly, 242—effects of light on them, 244—their circulating fluid, 269—destitute both of blood and of chyl-aqueous fluid, 270.
Actinophrys, assimilation in the, 220, 230.
Æschylus on the Sea, 285.
Aëration, mode of, in Annelids, 72—process of, in the Eolis, 113.
Agassiz on the Mactra, &c., 117—on the identity of the Medusa and the Polype, 339.
Age, effect of, on the brain tissue, 400.
Alcyonidium hirsutum, the, 354.
Alder and Hancock's Nudibranchiate Mollusca, 36—on the Pholas, 91 *note*.—their representations of the Eolids, 109—on the Doris, 112—on the respiration of the Eolis, 114.
Allmann, Professor, on the circulating fluid of the Polyzon, 275—his monograph on Polyzoa, 354.
Alternation of generations, 297. *See* Parthenogenesis.
Ammocete, metamorphosis of the, 253 *note*.
Amœba, constitution, &c. of the, 167—organisation of the, 220 *note*—digestion in it, 230.

Anemone, how determined to be an animal, 127 *et seq.*—voracity of the 133—its alleged paralysing power, 135—its reproductive system, 163—its structure, 164. *See also* Actinia.
——— Aurora, the, 83—its vitality, 21—identical with Venusta, 150.
——— bellis, 14—where found, 17—trial of its paralysing power, &c., 138—its tentacles, 165—effects of light on it, 244—nutriment derived from water by it, 271—observations on the supposed chyle-corpuscles in it, 274.
——— crassicornis, where found, 17—chiselling it out, 19—its voracity, 134—trial of its paralysing power, 138, 140—the thread capsules in it, 157—discovery of the ovaries in it, 170—at the Scilly Isles, 212—experiments on its supposed solvent fluid, 227—varieties of it, 253—variations in size in it, 270.
——— depressa, 129.
——— dianthus or Plumose, 129, 178—the thread capsules in it, 158—experiment on its urticating power, 160—Dr Wright on it, 178—reproduction in it, 180.
——— lacerata, reproduction in, 180, 182.
——— nivea, 253.
——— ornata, 292.
——— parasitica, oviparous, 149—at Jersey, 292.
——— rosea, reproduction in the, 180.
——— venusta, 83—its vitality, 21—and aurora, identity of, 150.
Anemones, importance of leaving their base uninjured, 20 *et seq.*—transmission of, by post, 82—general enthusiasm for them, 121—errors in books regarding them, 123 *et seq.*—their stinging power, 145—disproof of their alleged paralysing power, 146—their longevity and vitality *ib.*—edible, 147—viviparous, 148—uncertainty of colour as a distinction in them, 150—the so-called urticating

INDEX.

cells in them, 157. *See also* Actinia, &c.
Animal, anatomical distinctions of the, 133 *note*.
Animal envelope, universal importance of the, 165.
Animal heat, influence of, on maternal instinct, 251.
Animal and vegetable life, the mode of distinguishing between, 127 *et seq.*
Animal paradoxes, examples of, 55.
Animals, relations of their organism to light, 248—discovery of new, 262—the identification of, 295—and man, comparative perfection of sense in, 375.
Annelids, partiality of, for darkness, 22—the budding or gemmation of, 64—Bonnet's observations on, 66—peculiarities of the blood of, 70 *et seq.*—twofold respiration of, 72—entrance of water into the blood in, 117—digestion in, 231—effects of light on, 244—Parthenogenesis among, 308.
Anthea, its voracity, 136—trial of its paralysing power, 138, 140—changes of colour in it, 150—the thread capsules in it, 158—its tentacles, 165—self-division in it, 184.
—— cereus, where found, 17—experiments on the supposed solvent fluid of the, 227.
Antheas, motion in the, 135—at the Scilly Isles, 212, 242.
Anthropomorphism, tendencies to, in zoology, and danger of them, 587 *et seq.*
Aphides, reproduction in, 300 *et seq.*—Owen on it, 302—its ratio, *ib.*—researches of Huxley on the generation of, 315.
Aphrodita, blood of the, 70.
Aplysia, the, 23—superstitions regarding it, *ib.*—peculiarities of its ova, 260 *note*—the so-called eyes in the, 359—its nervous system, 392, 393. *See also* Sea-hare.
Apneusta, the order of, 115.
A priori conclusions, danger of, in Zoology, 387 *et seq.*
Aquarium at Regent's Park, the, 5.
Aristotle on the Anemones, 124.
Arran, colony of Dianthi at, 179.
Ascidian, the, 25, 96—an undescribed, 295—the embryo of the, 296.
Ascidians, want of motion in the, 61, 131—peculiarities of generation in the, 304.
Assimilation, distinction between digestion and, 219—chemical solution necessary to, 223—and digestion, not identical, 409.
Aucapitaine, M., on the Boring Molluscs, 87.
Auerbach, on the enveloping membrane, &c., 220 *note*.

Avicularium of the Corkscrew Coralline, the, 385.

BAER, VON, *see* Von Baer.
Baird, Dr, on the vitality of Molluscs, 358.
Basket, the collecting, for marine animals, 12.
Basques, peculiar custom among the, 250.
Beauty, charm of, 147.
Bee, Parthenogenesis in the, 308.
Beneden, *see* Van Beneden.
Bergmann and Leuckart on the blood, 223—on the convoluted bands, 277.
Bernard, Claude, on anatomy as a source of error, 162—on digestion, 225.
Berryn Narbor, a ramble to, 80.
Biliary secretions, provision for, in the Eolis, 110.
Biology, neglect of true induction in, 104.
Bird's-head process of the Corkscrew Coralline, the, 385.
Blessig on the structure of the retina, 363.
Blood, peculiarities of, in Annelids, 70 *et seq.*—entrance of water into the, 116—changes it undergoes before assimilation, 223—true, definition of, 269.
Bonnet, the experiments, &c. of, on Annelids, 66—on the reproduction of Aphides, 300.
Books, suitable, for marine studies, 35.
Boring Molluscs, the, 85—how they operate, 87—prevention of their ravages, 89.
Borlase's account of the Scilly Isles, 194—on the former state of the Scilly Isles, 203—historical notices from, 204.
Botryllus, the, 25.
Bovista giganteum, cell-formation in the, 332.
Brain, increase of fibre in the, with age, 400.
Braunton road and inn, Ilfracombe, 8.
Brightwell, Mr, on the Noctilucæ, 344.
Brittle-star, the, 254 *et seq.*
Budding or Gemmation of Annelids, 64. *See* Gemmation.
Bunodes Crassicornis, *see* Anemone.

CAILLAND, M., on the vitality of Molluscs, 358.
Calamary, the colour-specs on the, 99 *et seq.*
Calvados, the oyster-establishments at, 357.
Campanularian Polypes, Parthenogenesis among the, 311.
Capstone Parade, the, Ilfracombe, 29, 32.
Carp, the, a ruminating fish, 252.
Carpenter, Dr, on the Anemones, 131—on the digestion of the Polypes, 227—on Parthenogenesis, 310.

INDEX.

Carus, Victor, on the zoological wealth of the Scilly Isles, 241—on the convoluted bands, 277.
Cells, multiplication by division of, 331—by union of two similar, 334—and of two dissimilar, *ib*.
Cephalopods, the eyes of the, 359.
Cerianthus, peculiarity of sex in the, 174.
Chambercombe woods, 78.
Chamisso, discoveries of, on the generation of the Salpa, 303.
Chance-seeking, 353.
Chiaje, *see* Della Chiaje.
Chylaqueous fluid, what, 71—uses, &c. of it, 72—distinction between, and blood, 269.
Chyle-corpuscles, the supposed, in the Actiniæ, 271 *et seq*.
Ciliary action, is it distinguishable from life? 60.
Circulation, relations of, to respiration, 267.
Classification, difficulties of, 264.
Clavelinæ, the, 15—at Jersey, 291.
Cleopatra's pearl, the story of, 105.
Cockle, vitality of the, 355.
Colour, uncertainty of, as a specific distinction in the Anemones, 150.
Colour-specs, the, on the Loligo, 99 *et seq*.
Comatula, the, 216, 254.
Commerce, primitive state of, in the Scilly Isles, 206.
Commissariat, state of, in the Scilly Isles, 206.
Compound Ascidian, generation of the, 296.
Comte, Auguste, modification of Harvey's aphorism proposed by, 298.
Convoluted bands of the Anemones, what? 169—function of the, 277—not biliary or urinary organs, 278.
Contractility, existence of, apart from muscularity, 389—specialisation of, 406.
Coriaceous Anemone, the, 21. *See* Anemone crassicornis.
Corkscrew Coralline, the, 385.
Corpuscles, absence of, in the blood of Annelids, 70.
Coryne, reproduction in the, 180.
Couch, R. Q., on the sex of the Actiniæ, 178—experiments by, on the digestion of the Actiniæ, 229—on the circulating fluid of the Anemones, 272—on reproduction in the Hydra, 299—on reproduction in the Sertularian Polypes, 330.
Cowrie, the, 215.
Crabs, do the Anemones feed on? 138, 139.
Crassicornis, the, *see* Anemone.
Crinoidea, the, 216.
Crustacea, partiality of, for darkness, 22.
Currents, influence of Molluscs in inducing, 377.

Cuvier on the blood of the Annelids, 70.
Cydippe, the, 343, 346.

DAISY, the, 14—where found, 17. *See* Anemone bellis.
Dalzell, Sir John, on the strength, &c. of the Anemones, 137.
Daphnia, alleged paralysing power of the Hydra on the, 142.
Darkness, importance of, to many marine species, 22.
Decay, as a life-function, 62.
Della Chiaje on the colour-specs of the Loligo, 102 *note*—researches of, on the Anemones, 125—on the reproduction of the Actiniæ, 183—on nervous system, 391, 405 *notes*.
Dendronotus arborescens, the nervous system of the, 393 *note*.
Desert snail, vitality of the, 358.
Dianthus, the, *see* Anemone.
Dicquemare, the Abbé, 124—on the vitality of the Anemones, 21.
Differentiation, illustrations of, 167.
Digenesis, 298. *See* Parthenogenesis.
Digestion, organs of, in the Eolis, 109—what constitutes it, 219—distinction between it and assimilation, 220—simplest form of, 221—its successive forms, *ib. et seq*.—is mainly chemical, 223—what is included in it, 224—a special function, 225—experiments on the so-called, in the Actiniæ, 227 *et seq*.—advance of, in the animal series, 230—assimilation not identical with it, 409.
Dionæa muscipula, an example of a food-seizing plant, 130.
Dissections, how done, &c., 290.
Dog, the new-born, peculiarity in nervous system of, 395.
Dogfish, the nervous system of the, 394.
Dorids at Jersey, 291.
Doris, the, 47—the spawn of the, 24—respiration in it, 112—peculiarity of structure in foot of, 116—its development, 259—the shell rudiment in it, 261—the so-called eyes in it, 359—hearing in, 372—its nervous system, 390, 393.
——— lugubris, peculiarity of the eyes in the, 360.
——— tuberculata, the, 255.
Dragonfly, nerve-tissue of the larva of the, 401.
Draper, Professor, on animal development, 166—on the influence of light on plants, 246—his theory of vision, 365 *et seq*.
Dredging for marine animals, 92.
Dress, appropriate, for marine collecting, 18.
Druidical remains, supposed, in the Scilly Isles, 234.

EATING, defence of, 43.

420　　　　　　　　　INDEX.

Echinoderms, the, 254.
Ecker on the so-called muscles of the Hydra, 404.
Edwards, *see* Milne-Edwards.
Ehrenberg, the Polygastrica of, 230 *note*.
Elizabeth, Queen, fortification of the Scilly Isles by, 204.
Englishmen, characteristics of, abroad, 30—passion of, for the sea, 209.
Entomostraca, peculiarities of generation in the, 308.
Entozoa, alternation of generations among the, 304.
Envelope, importance of the, in animals, 165.
Enveloping membrane, supposed, in the Actinophrys, 220 and *note*.
Eolids, investigations into the respiration of the, 107—the thread capsules in the, 154, 156—their development, 259—at Jersey, 291.
Eolis, the, 27—description of it, 108—dissection of it, 109—its respiration, 111—not a Nudibranchiate, 112, 114—Owen and Siebold on it, 115—the shell rudiment in it, 261—the so-called eyes in it, 359.
——— alba, the, 215, 256.
——— landsbergii, 109.
——— papillosa, 108, 256.
——— pellucida or elegans, 108.
Eunice, peculiarity in the, 117.
Eye, alleged injury from the microscope to the, 39—a tactile organ, 361 *et seq*.
Eyes, the so-called, in Molluscs, 358.

FERUSSAC, Baron, on the insensibility of the lower animals, 349.
Filiferous capsules, *see* Thread capsules.
Fish, paradoxes and anomalies among, 250 *et seq*.
Fishes, eccentricities of, 54.
Fishermen, value of, to the marine hunter, 289—their stupidity, *ib*.
Fissiparous reproduction, what, 326.
Fleming, Professor, 146 *note*.
Flourens, M., on the gills of fish, 355.
Food, relations of life to, 231—parallel between, and knowledge, 232.
Forbes's Naked-eyed Medusæ and British Starfishes, 36—on the Crinoidea, 217—on the Brittle-star, 255, —on the Solen, 384.
Fresh-water Polypes, Trembley on the reproduction of the, 299.
Frog, the nervous system in the young, 394.
Function, relations between organ and, 62, 408 *et seq*.
Funke on the nervous system, 403.

GALL-FLY, peculiarity of sex in the, 308.
Ganglia, the, of the Doris, &c., 390 *et seq*.

Gem, the, 23, 25.
Gemmaceæ, haunts of the, 17—not migratory, 132—variations in size in the, 270.
Gemmation or budding of Annelids, the, 64—reproduction by, 299—identity of, with generation, 326.
Gemmiparous generation, what, 326.
Geneagenesis, 298. *See* Parthenogenesis.
Generation, gemmation identical with, 326—a mere form of growth, 328.
Germ cells and sperm cells, assumed necessity of both to seed, 327—reproduction by union of, 334.
Giant's Castle, Scilly, 201, 202.
Gills, difference between the, in Fishes and Molluscs, 255.
Glass jars, suitable, for Anemones, &c., 34.
Goethe's law of animal development, 167—his love of Anatomy, &c., 291—on growth and reproduction as identical, 331.
Goniaster, the, 254.
Gorey, village of, Jersey, 288.
Gosse, Mr, his collecting basket, 12—on the sea, 28—his Marine Zoology, 36—on the Pholas, 92—the works of, 125—on the edibility of the Anemones, 147—identity of his orange-disked and orange-tentacled Anemones, 150—new genus "Sagartia" proposed by, 158—on the Nymphon, 214—on the Corkscrew Coralline, 386.
Granite, predominance of, in the Scilly Isles, 240.
Granitic rocks, the, at the Scilly Isles, 213.
Granular nerves, previous observations on, 397.
Greeks, their love of the sea, 209.
Growth, reproduction only a form of, 331 *et seq*.
Gymnetrus, the, 25.

HAIME, M. Jules, memoir on the Cerianthus by, 174.
Hand, formation of the, 168.
Harvey, his aphorism as to reproduction modified, 298.
Harvey, Dr Alexander, on generation, 327.
Hearing, the so-called, of Molluscs, 272 *et seq*.
Helianthoid Zoophytes, reproduction in the, 181.
Hermit crab, habits, &c. of the, 49 *et seq*.—Swammerdamm on the, 258.
Hills, peculiarities of the, at Ilfracombe, 6—effects of, on works of man, 32.
Hippopotamus, the, 122.
Holland, M., memoirs of, on the Anemones, 126—anatomy of the Actiniæ by, 165—the ovaries of the Actiniæ

INDEX. 421

described by, 170 *note*, 171, 176—on the supposed solvent fluid of the Actiniæ, 228—on the convoluted bands, 277.
Homer on the sea, 212, 285.
Huxley, Professor, classification of the Sagitta by, 265—observations on it, 267—on Parthenogenesis, 304—on the generation of Aphides, 302 *note* 316.
Hydra, alleged paralysing power of the, 143—its organisation, 166—reproduction in it, 180—digestion in it, 230 *note*—Trembley on its reproduction, 299—peculiarities of reproduction in it, 328—its so-called muscles, 404.
Hydra tuba, reproduction in the, 180.
Hydractinia, the, 291.
Hydroid Polypes, reproduction in the, 180.

ILFRACOMBE, scenery of, 5—the town, 7—a day's marine hunting at, 16—the visitors to, 30—the Capstone Parade at, 32—a ramble near, and its scenery, 77—lanes, 78.
Implements, the requisite, for marine-animal collecting, 12.
Incubating fish, an, 251.
Infusoria, digestion in the, 230.
Ingenhousz, discovery of the influence of sunlight on plants by, 246.
Inman, Dr, on insensibility to pain in the lower animals, 351 *note*—on the nervous system of the Dendronotus, 393 *note*.
Insects, Parthenogenesis among, 308—signs of insensibility given by, 349—peculiarities of nervous tissue in, 401.
Invertebrata, position of the retina in, 368.

JELLY-FISH, preference of, for light, 22—the thread capsules in the, 154, 156—production of, by Polypes, 297.
Jersey, first impression of, 284—character of its scenery, 285—boyish reminiscences of it, 286 *et seq*.
Johnston's British Zoophytes, 36, 125—on the voracity of the Crassicornis, 133.
Jones, Rymer, his Animal Kingdom, 36—on the study of marine animals, 63—on the animal nature of the Anemone, 128—on their powers of motion, 132—on the seizing power of the Anemones, 138.

KIRKES and PAGET on nerve fibre, 396 *note*.
Knowledge, the food of the mind, 232.
Kölliker on the colour-specs of the Loligo, 102 *note*—on the sex of the Actiniæ, 172, 174, 175—observations of, on the Actinophrys, 220—on Parthenogenesis, 313—on the sensibility of the skin, 371.
Krohn, classification of the Sagitta by 265—on it, 267—on Parthenogenesis, 304—on the Noctilucæ, 344.
Kyber, experiment on the Aphis by, 325.

LAGENELLA REPENS, the, 262.
Lamp Polype, the, 254.
Lamprey, metamorphosis of the Ammocete into the, 253 *note*.
Landrail, the, 8.
Landsborough, Dr, on the Anemone, 131—on the stinging power of the Anemone, 136.
Lanes, the, in Scilly, 200.
Lankester, Dr, on Parthenogenesis in plants, 308.
Laomedea geniculata, reproduction in the, 297.
Lantern Hill, Ilfracombe, 32.
Leeuwenhoek, discovery of, respecting vertebrate nerve-tissue, 403 *note*.
Lernæa, peculiarities of the, 56.
Leuckhart on Parthenogenesis, 304, 305.
Leydig on the structure, &c. of the eye, 370—on the colouring matter in the ganglia of the Doris, &c., 391 *note*—on the so-called muscles of the hydra, 404.
Life, revelations of the mystery of, through marine studies, 56—is motion the index to? 61—definition of it, 62 *et seq*.—no special organs in the earliest forms of, 63—relations of, to food, 231—the relations of light to, 246 *et seq*.—the successive developments of, 259.
Light, indifference of Molluscs, &c. to, 22—alleged sensibility of the Anemones to, 128—effects of, on various marine species, 244—vital action of, *ib.*—its relation to organisation, 245—relations of, to the animal organism, 248—apparent indifference of Actiniæ, &c. to, 249.
Limax, the nervous system of the, 392, 393.
Lion, the man-eater, and the zoologist, 187.
Literature, greatness of, 278.
Locomotion, special organs of, 408.
Logic, inattention to, among zoologists, 104.
Loligo, the colour-specs on the, 99 *et seq*.
Lophius, the, 252.
Lucernaria, the, 254.
Luidia fragilissima, the, 255.

MADEIRA SNAIL, vitality of the, 358.
Mammalia, digestion in the, 231.
Man, perfection of the senses in, compared with animals, 375.
Marine animals, hunting for, at Ilfra-

combe, 11—eccentricities of formation, &c. among, 55—dredging for, 92—charms of, 42.
Marine species, various effects of light on, 244.
——— spider, the, 213.
——— zoology, attractions of, 4.
Maternal instinct, influence of animal heat on, 251.
Maury, Lieutenant, on the influence of Molluscs on oceanic currents, 377.
Medusæ, vitality of detached portions of the, 60—identity of, and the Polypes, 339—fishing for, 343—are they susceptible of pain? 347 *et seq*.
Meissner on granular nerves, 397 *et seq*.
Mermis, granular nerves in the, 397, 398, 399 *note*.
Mesembryanthemum, the, 13—left dry by the ebb, 132. *See also* Anemone.
Microscope, employment of the, 36—its value and uses, 38—errors concerning it, 39—considered as a new sense, 57.
Milne-Edwards on the blood of Annelids, 70—on alternation of generations in the Ascidians, 304—on the oyster, 357—on the relation of function to organ, 408.
Mind, effect of the study of nature on the, 57, 412.
Moleschott, the experiments of, on the relations of light and life, 249.
Molgula arenosa, 96.
Molluscs, general partiality of, for darkness, 22—habitations of, and those of men, 33—the shells of, 261—Parthenogenesis among, 308—insensibility of, to pain, 349—powers of endurance in, 355—the so-called eyes of, 358, *et seq*., 370 *et seq*.—the bearing of, 372 *et seq*.—their shells, 375 *et seq*.—nervous system of, 390 *et seq*.
Mormon preacher, &c., a, 103.
Morte stone, the, at Ilfracombe, 83.
Moths, Parthenogenesis among, 307.
Motion, relations between, and life, 61.
Mulder on the green of leaves, 245.
Müller, Memoir by, on the sounds made by fish, 253.
Müller and Gegenbaur, observations on granular nerves by, 397.
Muscle and nerve, analogous development of, 401.
Muscular tissue, assumption as to, in the Annelids, 73.
Muscularity, contractility apart from, 389.
Myriana, gemmation of the, 64.

Nais, the gemmation of the, 64—alleged paralysing power of the Hydra on it, 142, 144.
Nais proboscidea, gemmation of the, 68.
Naked-eyed Medusæ, the, 343.
Nature, effect of study of, on the mind, 57, 412.

Nemertes Borlasia, the, 236.
Nereis, peculiarity in the, 117.
Nérine, extraordinary adventures of the crew of the, 236.
Nerve and muscle, analogous development of, 401—stages of development in different orders, 402.
Nerves, sensations where they are invisible, 371—sensibility manifested without, 404.
Nerve-fibres, and cells, alleged distinction between, 392—supposed dependence of nerve-action on, 394 *et seq*. —function of the, 400.
Nerve-force, and action, relations of, to nerve-fibres, 394—not conducted by fibres alone, 399.
Nervous system, connection of sensibility to pain with the, 350—sensibility apart from the, 389.
Nervous tissue, assumptions as to, 74—is it ever diffused? *ib.*—constitution of, 400.
Noctilucæ, the, 344, 345.
Nudibranchiates, respiration in the, 112—structure, &c. of the eye in, 352 *et seq*.
Nutrition, as a function of life, 61, 62.
Nymphon gracile, the, 213.

Observation, value of, to the Naturalist, 36.
Ocean-currents, influence of Molluscs on, 377.
Ophiolium, rapid death of, out of water, 355.
Orange-tentacled and disked Anemones, the, 83—their identity, 150.
Orbigny on the colour-specs of the Loligo, 99.
Organ, relations of, to function, 62, 408 *et seq*.
Organic labour, progressive division of, 407.
Organisation, relations of light to, 246.
Ova of the Eolis, &c., peculiarities of the, 259, 260.
Ovaries, discovery of, in the Actiniæ, 170.
Owen, Professor, on the Boring Molluscs, 88—on the Eolis, 115—on the paralysing power of the Hydra, 142—on the urticating cells, 156—on the digestive organs of the Pleurobranchus, 258—his "Parthenogenesis," 298—on the generation of the Aphis, 302—his lectures on Parthenogenesis, 305—his theory of Parthenogenesis, 318 *et seq*.—examination of it, 321 *et seq*—on the identity of gemmation and generation, 326—on pain in the lower animals, 349—on the eye of invertebrates, 368—on the colouring matter in the ganglia of the Doris, 391.
Oyster, preservation of the, out of water, 357.

INDEX. 423

PAGURUS, or Hermit crab, habits, &c. of the, 49 *et seq.*

Pain, is it felt by the Medusæ? 347 *et seq.*—the manifestations of, in man, and error in transferring these to the lower animals, 348—shrinking, cries, &c., not necessarily signs of, 349—its connection with the nervous system, 350—sensibility to, a specialisation of the general sensibility, 351—differences of it, 352.

Paramecium, rapidity of increase of the, 332.

Parthenogenesis, what, 297 *et seq.*—history of its discovery, &c., 298 *et seq.*—statement of, 309—new facts in it, 310 *et seq.*—theories of, 317—author's theory of, 338 *et seq.*

Pedicellaria of the Starfish, the, 387.

Pedicellina, the, 263—is viviparous as well as oviparous, &c., 264.

Peltier, M., observations of, on Annelids, 66.

Penzance lodging-house, a, 193.

Perch, sexual peculiarity of the, 177.

Perca scandens, the, 252.

Pholas dactylus, the, 85, 87, 90.

Phosphorescence of the sea, source of the, 344, 345.

Phyllirhoë, granular nerves in the, 397.

Physalia, peculiarities of the, 55.

Pipefish, the, 215—its nervous system, 394. *See* Syngnathus.

Planariæ, the, 112—the thread capsules in, 154, 156.

Plants, sensitive to light, 128—examples of their seizing insect-food, 130—locomotion in, 131—relations of light to, 246 *et seq.*—Parthenogenesis among, 308.

Pleurobranchæa, the, 372.

Pleurobranchus, the, 256—its digestive organs, 257—the shell rudiment in, 261—the so-called eyes in, 359—hearing in, 372—its nervous system, 390, 393.

Plumatella, the, 215.

Plumose Anemone, *see* Anemone Dianthus.

Plumularia myriophyllum, Parthenogenesis in the, 311, 313.

Plumularian Polypes, Parthenogenesis among, 311.

Poets, the, on the sea, 285.

Polygastrica of Ehrenberg, the, 230 *note.*

Polynoë, blood of the, 70.

Polypes, the alleged paralysing power of the, 135—the thread capsules in the, 154, 156—do they truly digest, 226—digestion in, 230—peculiar reproduction in, 297—production of jelly-fish by, *ib. et seq.*—Parthenogenesis among, 308—identity of, and the Medusa, 339—sensibility manifested by, 404.

Polypes, the fresh-water, their paralysing power disproved, 145—hermaphrodite, 177.

Polypus eudendrium, peculiarities of generation in, 313.

Polyzoa, the, 215—Allmann on the fluids of the, 275—molluscan nature of, 354.

Polyzoon, a new, 354.

Pond, a, 74.

Pond-mussels and snails, vitality of the, 358.

Post, live animals sent by, 82.

Pratt, Mrs, on the stinging power of the Anemone, 136.

Priestley, discovery of the expiration of oxygen by plants, by, 245.

Protococcus nivalis, reproduction in the, 332.

Protoplasma, the, 331.

Public whipping in Jersey, 287.

Pycnogonidæ, the, 112.

QUATREFAGES, M., and the microscope, 38—on the Annelids, 117, 118—name given to Parthenogenesis by, 298—his views on it, 306, 322 *et seq.*—on Owen's theory of it, 322—on gemmation, 329—on the Noctilucæ, 344.

RAPP, his work on the Anemones, 125—on the sensibility of the Anemones to light, 129—on the sex of the Actiniæ, 173, 174.

Ratiké on the Annelids, 117.

Razor-fish or Solen, the, 380 *et seq.* —*See* Solen.

Réaumur on the Anemones, 124—on the motion of the Anemones, 132—the experiments of, on digestion, 228.

Red Nose, the, 87.

Reproduction as a life-function, 61, 62, —by gemmation, on, 64—system of, in the Anemone, 163—recognised forms of, 326—and growth, essential identity of, 331—processes of, by division of cells, 332 *et seq.*—by union of two similar cells, 334—by that of two dissimilar cells, *ib.*

Respiration, twofold, in Annelids, 72—of the Eolids, investigation into the, 107, 111 *et seq.*—influence of light on, 248—relations of, to circulation, 268—specialisation of, 406.

Retina, error as to images being formed on the, 361—its structure, 362—insensible to light, 364—its position in Invertebrates, 368.

Ribbon-fish, the, 25, 215.

Robin on the colour-specs of the Loligo, 102 *note.*

Rocks, the, at Ilfracombe, 9.

Rock-basins, the supposed Druidical, 235.

Rondelet on the Anemones, 124.

SABELLÆ, the, 69—blood of the, 70.

Sagartia, proposed new genus of, 158.

Sagitta Mariana, difficulties of classifying the, 265—description of it, *ib.*—its want of a vascular system, 267.
St Helier's, town of, 286, 288.
St Mary's Isle, one of the Scillies, 195, 199—Sound, 197.
Salpa, peculiarities of generation in the, 303.
Sars, discoveries of, in Parthenogenesis, 304.
Sarsia prolifera, reproduction in the, 183.
Saxicava rugosa, the, 87.
Scepticism, importance of, in Biology, 105—necessity of, in scientific observation, 162.
Schnarda on chyle-corpuscles in the Actiniæ, 274.
Science, delights of, 45.
Scilly Isles, selection of, for sea-side studies, 191—voyage and landing at, 195 *et seq.*—first sight of, 197—description of them, 199—origin of their name, 203—historical notices of them, 204—the population, 205—geological conformation of the, 240—their climate, 242—departure from the, 283.
Sea, attractions of the, 4—aspect of the, its shore, &c. at Ilfracombe, 9—sources of its interest, 28—passion for it, 98—passion of Englishmen for it, 209—the poets on it, 285—source of it phosphorescence. 345.
Sea Anemones, the digestion of the, 219, 226 *et seq.* See Anemones.
Sea-hare, the, 23—superstitions regarding it, *ib.*—the shell rudiment in it, 261—at Jersey, 291. *See also* Aplysia.
Sea-mouse or Aphrodita, blood of the, 70.
Sea-scene, a, at Ilfracombe, 27.
Sea-sickness, 95, 97, 197.
Sea-slugs, *see* Eolids.
Sea-urchin, the, 254.
Seeing, *see* Vision.
Segond on nerve-fibre, 396 *note.*
Sennebier, discovery of the decomposition of carbonic acid by plants by, 246.
Sensations without visible nerves, on, 371.
Sensibility, that to pain a specialisation of, 351—existence of, apart from nervous system, 389 — manifested without nerves, 404—definition of, and whence derived, 406—specialisation of, *ib. et seq.*—does not necessarily imply senses, 409.
Serpulæ, the, 69—abundance of, 13.
Sertularian Polypes, reproduction in the, 330.
Seven Tors, the, Ilfracombe, 8.
Sex, determination of, in the Anemones, 172.
Shelled and naked Molluscs, the, 261.

Shells of Molluscs, the, 261, 375.
Shovel, Sir Clondesley, scene of the shipwreck of, 233.
Siebold, *see* Von Siebold.
Silk-worm, peculiarities of generation in the, 308.
Skin, universal sensibility of the, 371.
Sleep, influence of, on respiration, 248.
Slug, insensibility of the, to pain, 349.
Smooth Anemone, the, 13 — where found, 17.
Snails, vitality of various, 358.
Society at the sea-side, on, 30.
Solen, irritability of the, after death, 349—how to catch the, 380—salting their tails, 383—the nervous system of the, 392, 393.
Solitary Ascidians, the, at Jersey, 291.
Solubility, necessary to assimilation, 223.
Solvent fluid, the supposed, in the Actiniæ, 227.
Sperm-cells, assumed necessity of, to growth, 327—and germ-cells, reproduction by union of, 334.
Spermatozoa, present with ova in the Anemones, 173 *et seq.*
Spix, researches of, on the Anemones, 125.
Sponges, the circulation of the, 269.
Star Castle, Scilly Isles, 198, 204.
Starfish, a suicidal, 258—the Pedicellaria of the, 387.
Steenstrup on Parthenogenesis, 298, 305, 339—his theory of it, 317.
Stilling on the nervous system, 388 *note*—on the nerve-sheath, 402 *note.*
Stinging power of the Anemones, the alleged, 135.
Summer, delights of, 341.
Sunlight, influence of, on plants, 246.
Swammerdamm on the blood of the Annelids, 70—on the Hermit crab, 258.
Syllis, gemmation of the, 64—Parthenogenesis among, 308.
Syngnathus anguineus, the, 215—peculiarities of the, 250—the nervous system of the, 394.

Tadpole, a Molluscan, 296.
Teale, Mr, the ovaries of the Actiniæ described by, 170 *note*, 171.
Temperature, influence of, on generation, 333—vision dependent on, 365 *et seq.*
Ten Thousand, the, and the Sea, 209.
Tenby, compared with Ilfracombe, 5—sands, 75.
Tentacles of Annelids, &c., Williams on the, 72.
Terebella, the, 26—vitality of its separated tentacles, 59—gemmation of it, 64—no muscular tissue in it, 75.
Terebella nebulosa, the, 63, 64.
Teredo navalis, the, 86 *et seq.*

Thread capsules, are they organs of urtication, 154 et seq.
Todd, Dr, on interrupted continuity of nerve-fibre and its effects, 395.
Trawling, description of, 293.
Trembley on the fresh-water Polypes, 144—experiment on the Hydra by, 230 note—on the reproduction of the Polype, 299.
Tresco, climate of, 243.
Triton, irritability of the, after death, 350—its nervous system, 394.
Troglodytes, haunts of the, 17—not migratory, 132.
Tubicolæ, gemmation among the, 65.
Tubularia parasitica, 354.
Typical forms, interest of, 36.

URTICATING cells, the alleged, in the Polypes, 154 et seq.

VALLEYS, want of, at Ilfracombe, 6.
Van Beneden, name given to Parthenogenesis by, 298—his views on it, 306.
Vegetable and Animal life, the mode of distinguishing between, 127 et seq. See Plants.
Venus flytrap, see Dionæa Muscipula.
Vision, the process of, 360 et seq.
Vitality, peculiarities of, 59.
Vivisection, defence of, 347.
Von Baer, his law of animal development, 167.
Von Siebold on the Eolis, 116—on the Annelids, 117—on the paralysing power of the Hydra, 142—classification of the Sagitta by, 265—views of, on Parthenogenesis, 306, 322—on the eye of Invertebrates, 369—on the colour of the ganglia of the Doris, &c., 390.
Vorticellæ, the parasitic, 56.

WAGNER on the sex of the Actiniæ, 173.
Water, entrance of, into the blood, 116.
Water-newt or Triton, see Triton.
Whelk, the, and the Hermit crab, 53.
Wild-flowers, varieties of, in Scilly, 200.
Will, ready ascription of, by zoologists, to the lower animals, 100.
Williams, Dr T., on the Annelids, 67 et seq.—his views on the chylaqueous fluid, 71—on the tentacles of the Annelids, 72 et seq.—definition of a nutritive fluid by, 269—on that of the Anemone, 271, 273.
Wolff on the flower and fruit, 324, 327.
Wollaston, Mr, on the vitality of Molluscs, 358.
Woodward, Mr, on the Molluscs, 357.
Worms, see Annelids.
Wright, Dr Strethill on the dianthus, 178—the Actinia ornata described by, 292.

YEAST, reproduction in, 333.

ZOOLOGISTS, inattention to logic among, 102, 104—increasing enthusiasm of the, 188.
Zoology, necessity of scepticism in, 102—danger of *a priori* conclusions in, 387 et seq.
Zoophytes, partiality of, for darkness, 22.

DESCRIPTION OF THE PLATES.

FRONTISPIECE.

Fig. 1. Group of *Plumatella*. The figure to the right is of natural size; that to the left is magnified, and shows one animal contracted and two expanded, with their ciliated tentacles in activity. The group is fixed on a stem of sea-weed. (Copied from Milne Edwards.)

... 2. A *Cydippe* slightly magnified; the streamers contracted.

... 3. A *Noctiluca Miliaris* greatly magnified. The circular spaces are vacuolæ, filled with granules of food; these are only temporary, and are sometimes not visible at all. (Copied from Quatrefages.)

... 4. Group of *Clavellina*, about three times the natural size, rising from a creeping stem. The two orifices open and shut rhythmically to suck in and eject water. The lace-work structure is richly ciliated, and is supposed to represent the gills of the animal. (Copied from Mr Gosse.)

PLATE II.

Fig. 1. An *Eolis coronata*.

... 2. A *Doris Johnstoni*. The two horn-like processes on the head are the antennæ which retract, when the animal is touched, or removed from the water; the coronet at the other end is the circle of branchial plumes.

3. A *Pholas dactylus*, or rather the shell of one. (Figs. 1–3 copied from Woodward's Mollusca.

... 4. An *Aplysia*. The gills are seen under the fold of the skin, to the right. (Copied from Milne Edwards.)

5. A *Teredo Norvegica* removed from its shell. (Copied from Woodward's Mollusca.)

PLATE III.

Fig. 1. A diagram intended to exhibit the structure of an *Actinia*, but which must not be taken as strictly accurate. *a* is the base, *b* the disc, *c* the tentacles, *d* the mouth, *e* the stomach, *f g* and *k* the septa dividing the cavity into chambers, on the free surface of which are the grape-like masses of ovary, and in front of the ovary the convoluted bands. (Copied from Sharpey.)

... 2. Represents a section of the *Actinia* as seen in reality. The convoluted bands are seen covering the ovary, and terminating on the inner wall of the stomach. (Copied from M. Hollard.)

... 3. A magnified view of one convoluted band and ovary, with the mesentery between them, as they appear when unfolded on the glass slide, previous to examination under the microscope. (M. Hollard.)

... 4. Three *Thread-capsules* from an Actinia, greatly magnified; one with the thread still inside, the other two after the ejection.

... 5. Vertical section of the *Retina* of a Perch, showing the relative positions of the pigment layer, the rods and cones, the granules, fibres, and cells. (Copied from H. Müller.)

PLATE IV.

Fig. 1. *Campanularia.*—The figure to the right is of the natural size; that to the left is a branch greatly magnified. The Polype is seen, flower-like, expanding itself at the summit. An ovarian capsule, to the right, contains Medusæ in various stages of development. (Copied from Van Beneden.)

... 2. *Plumularia myriophyllum*, natural size. Five of the branches are seen developed into ovarian capsules.

... 3. Represents the spontaneous subdivision of one worm into two. The head of the second worm is seen formed at the segment where the tail ought to be. (Copied from Milne Edwards.)

PLATE V.

Fig. 1. *Lamproglena pulchella*, one of the crustacean Epizoa, greatly magnified, and seen from below. The figure is copied from Nordmann, with the addition of the ovisacs, not represented in his plate, but which Burmeister, whose copy I possess, has drawn in pencil. The *Lamproglena* is found in the gills of a fish (*Cyprinus Jeses*).

... 2. *Sagitta Mariana* greatly magnified; natural size quarter of an inch. I believe this to be a new species, and I have named it the *Mariana*. The ova are seen on each side of the alimentary canal; and in the lower half of the body are the cells containing spermatozoa, which issue from the orifices in the fin near the tail.

FIG. 3. *Nymphon gracile* (one of the *Pycnogonida*); natural size. The three circles from which the legs spring convey an inaccurate idea of the reality; they are not cavities, but enlargements of the trunk.

... 4. The same animal in the egg, just before it is hatched; greatly magnified.

PLATE VI.

FIG. 1. *Syngnathus anguineus* (Pipe-fish), about half the natural size.
... 2. A *Comatula rosacea*. (Copied from Forbes.)
...3-10. Development of *Strongylus auricularis*, one of the Entozoa. Fig. 3 shows the primary germ-cell, surrounded by the yolk; at 4, a division has taken place; 5, a subdivision; 6, still further subdivision; 7, the repeated subdivisions have resulted in what is called the "mulberry mass," out of which the embryo is gradually envolved as in figs. 8, 9, 10. (Copied from Bagge.)

PLATE VII.

FIG. 1. *Terebella nebulosa* removed from its tube. The tentacles are both longer and more numerous than here represented; and the gill-tufts at the side of the head are more apparent. (Copied from Rymer Jones.)
... 2. An ideal representation of the brain and spinal chord, showing how the nerve-fibres are theoretically supposed to communicate with the nerve-cells of the grey matter of the chord, and thence pass up to the brain, terminating in the cells of the grey matter of the convolutions. (Copied from Leydig.)
... 3. A *Pleurobranchus*, natural size, seen from above.
   4. A stem of the *Corkscrew Coralline*, natural size.

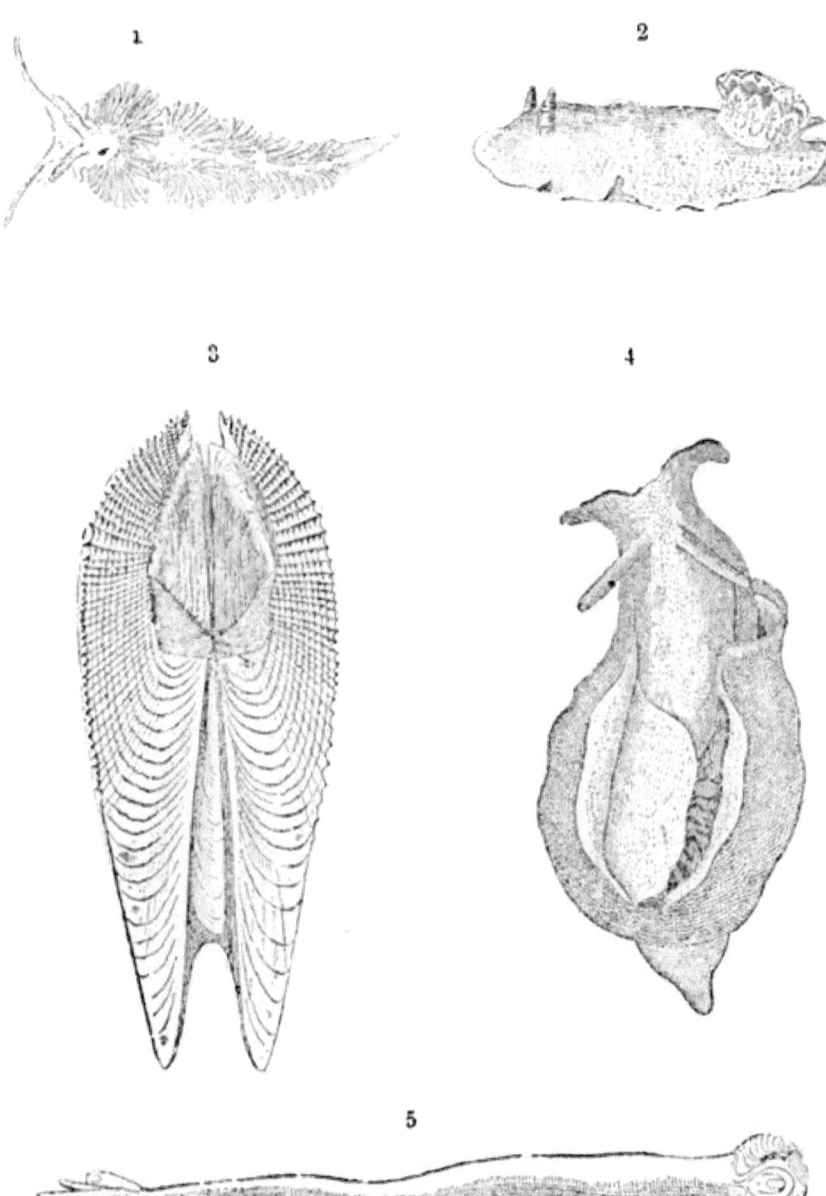

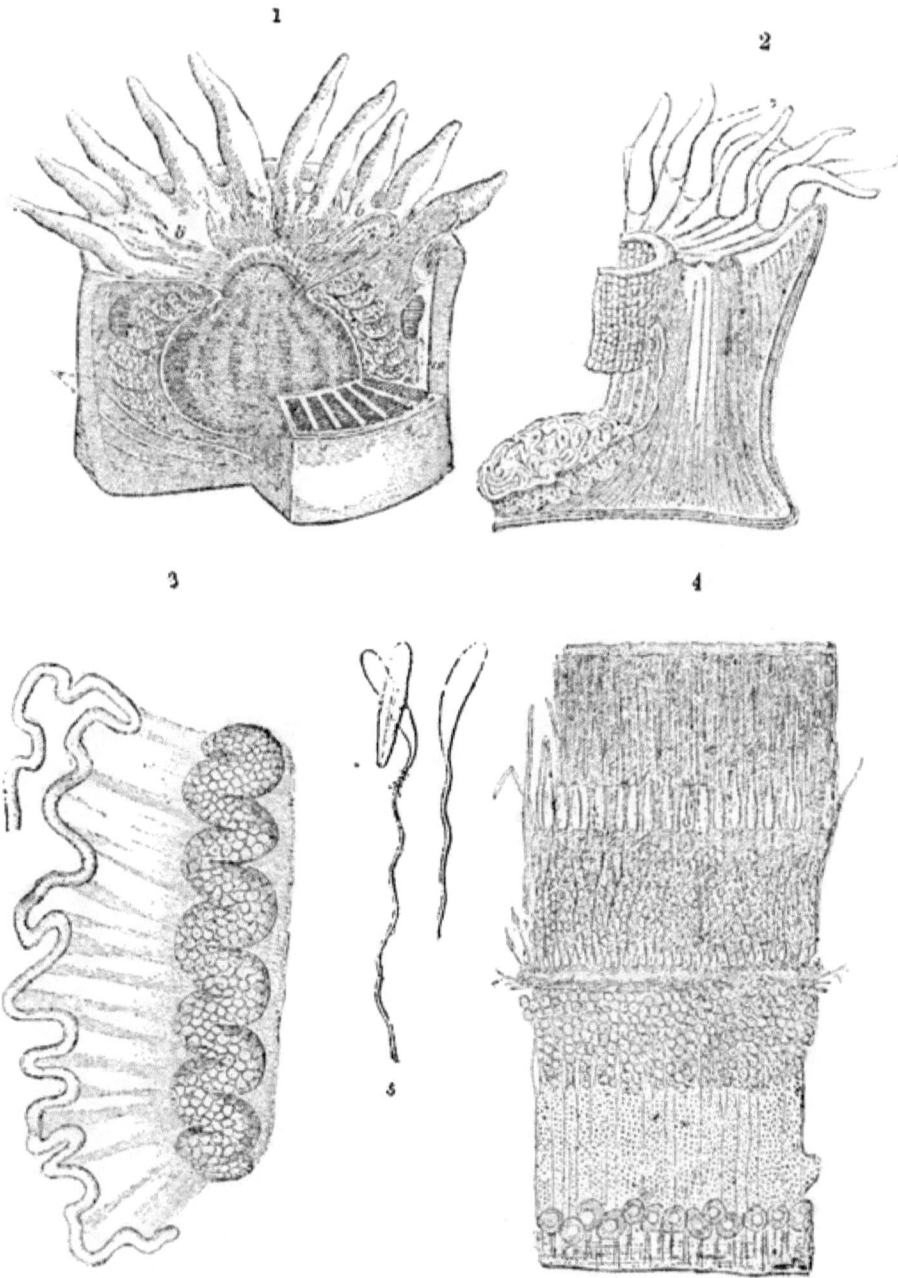

PLATE IV

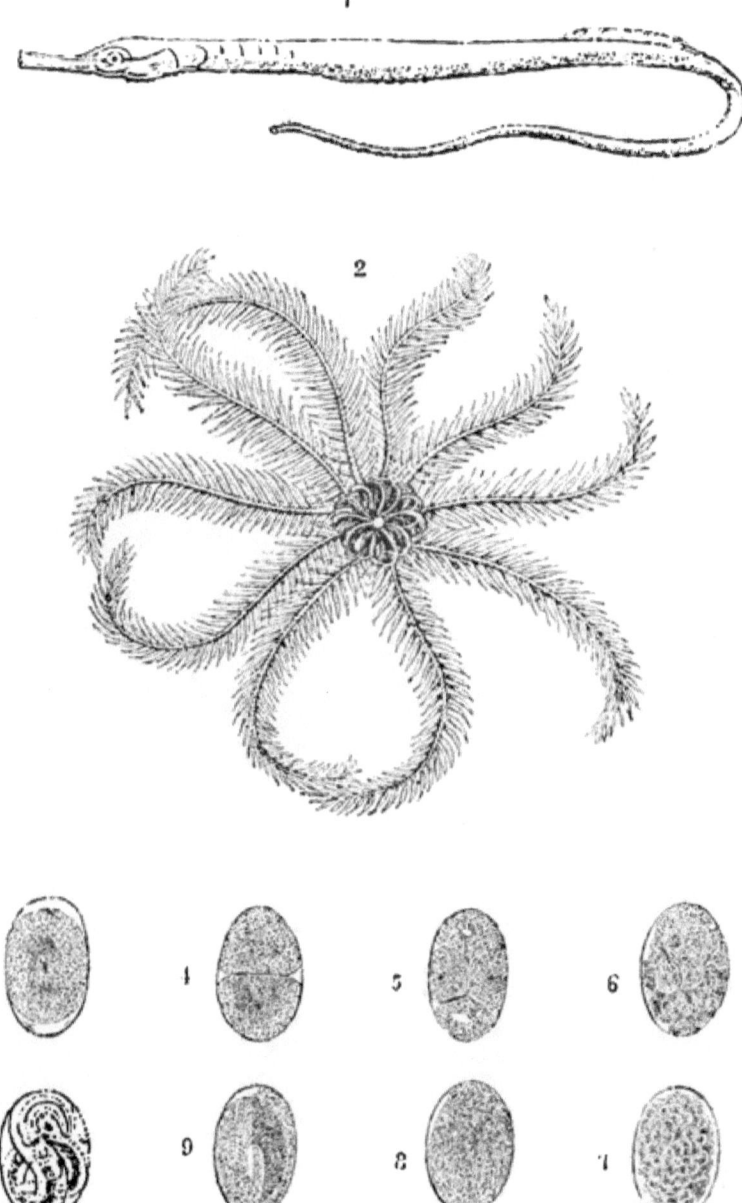

PLATE VII

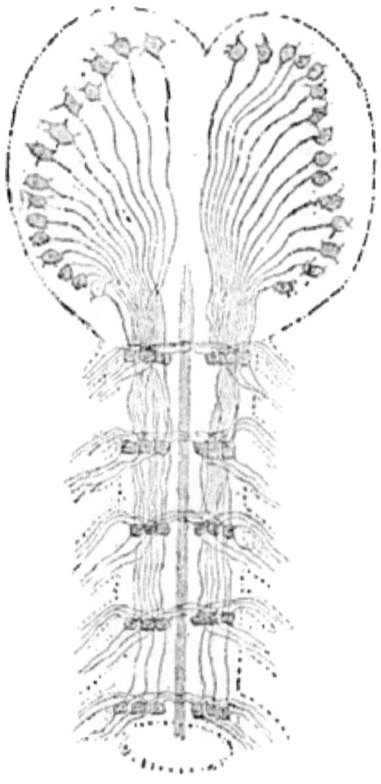

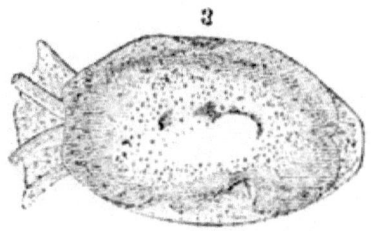

MESSRS BLACKWOOD'S
PUBLICATIONS.

Complete in Two Volumes,
THE PHYSIOLOGY OF COMMON LIFE.
By George Henry Lewes,
Author of "Sea-Side Studies," "Life of Goethe."
Illustrated with numerous Engravings on Wood, price 12s.

"Mr Lewes brings the tact of a man of taste to the work of a drudger, and tells, with the ready wit of a skilled writer, what has been learned by many tedious processes of study. . . . 'The Physiology of Common Life' is a good subject in good hands."—*Examiner.*

In Two Volumes,
THE CHEMISTRY OF COMMON LIFE.
By Professor Johnston.
A NEW EDITION,
Edited by G. H. Lewes,
Author of "Physiology of Common Life."
Illustrated with numerous Engravings, price 11s. 6d.

"The most practically-useful volume which has ever appeared upon subjects with which every man ought to be acquainted. All should read it and refer to it, until the knowledge it imparts is as familiar as are already the matters of which it treats.'—*Tait's Magazine.*

FOURTH EDITION.
INTRODUCTORY TEXT-BOOK OF GEOLOGY.
By David Page, F G S.
Price One Shilling and Sixpence, bound in cloth.

BY THE SAME AUTHOR.
ADVANCED TEXT-BOOK OF GEOLOGY,
DESCRIPTIVE AND INDUSTRIAL.
A New Edition. Crown Octavo, with Illustrations, price 6s.

"We have read every word of it, with care and with delight, never hesitating as to its meaning, never detecting the omission of anything needful in a popular and succinct exposition of a rich and varied subject."—*Leader.*

BY THE SAME AUTHOR.
HANDBOOK OF GEOLOGICAL TERMS & GEOLOGY.
In Crown Octavo, price 6s.

"But Mr Page's work is very much more than simply a translation of the language of Geology into plain English; it is a Dictionary, in which not only the meaning of the words is given, but also a clear and concise account of all that is most remarkable and worth knowing in the objects which the words are designed to express. In doing this he has chiefly kept in view the requirements of the general reader, but at the same time adding such details as will render the volume an acceptable Handbook to the student and professed geologist."—*The Press.*

THE PHYSICAL ATLAS OF NATURAL PHENOMENA.

By Alex. Keith Johnston, F R S E., &c.
Geographer to the Queen.

A New and Enlarged Edition.

Imperial Folio, price L.12, 12s. half-bound morocco.

"The first edition of the Physical Atlas was a volume of which we were entitled, as a nation, to be proud. It supplied with a rare perfectness the want not of this country only, and its rapid sale was an inevitable consequence. By making the second edition, as this is, so clearly an improvement on the first, Mr Johnston secures for his Atlas a pre-eminence that it is not likely to lose during the lifetime of the present generation. It is indeed a work of magnificent range and completeness."—*Examiner.*

———o———

THE PHYSICAL ATLAS.

Reduced From the Imperial Folio.

By Alex. Keith, Johnston, F R S E., &c.

This Edition contains Twenty-five Maps, including a Palæontological and Geological Map of the British Islands, with descriptive Letterpress, and a very copious Index. In Imperial Quarto, half-bound morocco, price L.2, 12s. 6d.

———o———

GEOLOGICAL MAP OF SCOTLAND.

From the most recent authorities and personal observation.

By James Nicol, F R S E., &c.,
Professor of Natural History, Aberdeen;

AND

Alexander Keith Johnston, F R S E., &c.

Mounted on cloth, in cloth case, price 21s.

———o———

A MAP OF THE GEOLOGY OF EUROPE.

By Sir Roderick I. Murchison, D C L., M A, F R S.

AND

James Nicol, F R S E, F G S,
Professor of Natural History, Aberdeen.

On Four Sheets Imperial Folio, price L.3, 3s.; or L.3, 10s., mounted on cloth, in a Case.

———o———

AN ATLAS OF ASTRONOMY.

A complete Series of Illustrations of the Heavenly Bodies, drawn with the greatest care, from Original and Authentic Documents.

By Alex. Keith Johnston, F R S E, &c.,
Geographer in Ordinary to Her Majesty for Scotland;
Author of "The Physical Atlas," &c.

Edited by J. R. Hind, F.R.A.S.

Imperial Quarto, half-bound morocco, price 21s.

"To say that Mr Hind's 'Atlas' is the best thing of the kind is not enough, it has no competitor."—*Athenæum.*

JOHNSTON'S PHYSICAL ATLAS.

The following Maps may be had separately:—

THE DISTRIBUTION OF MARINE LIFE, illustrated chiefly by Fishes, Molluscs, and Radiata; with Map of the Colonisation of the British Seas, illustrated by Mollusca and Radiata; now first laid down by PROFESSOR EDWARD FORBES, F.R.S., &c. &c.; with Four Pages of Letterpress, Descriptive and Explanatory. Price 10s. 6d.

GEOLOGICAL AND PALÆONTOLOGICAL MAP OF THE BRITISH ISLANDS. In Two Sheets, including Tables of the Fossils of the different Epochs, &c. &c., from the Sketches and Notes of Professor EDWARD FORBES, &c.; with Eight Pages of Illustrative and Explanatory Letterpress, by PROFESSOR FORBES. Price 21s.

THE GEOLOGICAL STRUCTURE OF THE GLOBE, according to AMI BOUE, with Additions and Corrections to 1855, by A. K. JOHNSTON, F.R.S.E., &c.; with Eight Pages of Notes and Illustrations by J. P. NICHOL, LL.D., Professor of Astronomy in the University of Glasgow. Price 15s.

GEOLOGICAL MAP OF EUROPE, exhibiting the different Systems of Rocks according to the most recent researches and inedited materials. By SIR RODERICK IMPEY MURCHISON, D.C.L., F.R.S., &c. &c.; and JAMES NICOL, F.R.S.E., &c., Professor of Natural History, University of Aberdeen. With Four Pages of Descriptive and Illustrative Letterpress, complete Index of European Rocks, &c. Price 10s. 6d.

GEOLOGICAL MAP OF THE UNITED STATES AND BRITISH NORTH AMERICA, constructed from the most recent documents and unpublished materials, by PROFESSOR H. D. ROGERS, Boston, U.S.; with Six Pages of Descriptive and Illustrative Notes. Price 10s. 6d.

COMPARATIVE VIEWS OF REMARKABLE GEOLOGICAL PHENOMENA. By A. K. JOHNSTON, F.R.S.E., &c. Including Plans and Views of Vesuvius and Ætna—of the Island and Peak of Teneriffe—of Arthur's Seat—of South Keeling Islands, Grahame and Ascension Islands—of Crater of Gedee—of Volcanoes of Pichincha and Antisana, &c.; with Two Pages of Explanatory Letterpress. Price 7s. 6d.

THE PHYSICAL CHART OF THE ATLANTIC OCEAN, showing the Form and Direction of the Currents—the Distribution of Heat at the Surface—Navigation and Trade Routes—Banks, Rocks, &c. By A. K. JOHNSTON, F.R.S.E., &c.; with Six Illustrative Pages of Letterpress—including a Chart of the Basin, and a Vertical Section of the Atlantic, by LIEUT. MAURY; a general Sketch-Chart of the Oceanic Currents, and a CHART of the ARCTIC BASIN, with Description, by PROFESSOR H. D. ROGERS. Price 15s.

PHYSICAL CHART OF THE INDIAN OCEAN, showing the Temperature of the Water, the Currents of the Air and Ocean, Directions of the Wind, Districts of Hurricanes, Regions of Monsoons and Tyfoons, Trade Routes, &c. &c.; with Two Pages of Descriptive and Illustrative Notes. By A. K. JOHNSTON, F.R.S.E., &c. Price 7s. 6d.

PHYSICAL CHART OF THE PACIFIC OCEAN, showing the Currents and Temperature of the Ocean, the Trade Routes, &c. &c.; with Two Pages of Descriptive Letterpress. By A. K. JOHNSTON, F.R.S.E. Price 7s. 6d.

TIDAL CHART OF THE BRITISH SEAS, showing the Progress of the Wave of High Water, the Hour of High Water in Greenwich Time at New and Full Moon, and the Depth of the Sea. Constructed under the Direction of J. SCOTT RUSSELL, Esq., F.R.S.E., by A. K. JOHNSTON, F.R.S.E.; with Two Pages of Explanatory Notes, including a Tidal Chart of the World, by J. S. RUSSELL. Price 7s. 6d.

THE GEOLOGY OF PENNSYLVANIA.

A Government Survey; with a General View of the GEOLOGY OF THE UNITED STATES, Essays on the Coal-Formation and its Fossils, and a Description of the Coal-Fields of North America and Great Britain.

By Professor Henry Darwin Rogers, F R S, F G S,
Professor of Natural History in the University of Glasgow.

With Seven large Maps, and numerous Illustrations engraved on Copper and on Wood. In three Volumes, Royal Quarto, £8, 8s.

---o---

THE LECTURES OF SIR W. HAMILTON, BART.

ON METAPHYSICS AND LOGIC;

With Notes from Original Materials, and Appendix containing the Author's Latest Development of his New Logical Theory.

**Edited by the Rev. H. L. Mansel, B D, Oxford;
and John Veitch, M A, Edinburgh.**

In Four Volumes Octavo, price £2, 8s.

Vols. III. and IV. in a few weeks.

---o---

INSTITUTES OF METAPHYSIC.

THE THEORY OF KNOWING AND BEING.

By James F. Ferrier, A B, Oxon.
Professor of Moral Philosophy and Political Economy, St Andrews.

Second Edition. Crown Octavo, price 10s. 6d.

---o---

THORNDALE; OR, THE CONFLICT OF OPINIONS,

By William Smith,
Author of "Athelwold: a Drama," "A Discourse on Ethics," &c.

"Sleeps the future, like a snake enrolled,
Coil within coil."—WORDSWORTH.

Second Edition, Crown Octavo, price 10s.

---o---

THE EIGHTEEN CHRISTIAN CENTURIES.

By the Rev. James White.

Third Edition, with Analytical Table of Contents, and a Copious Index.

Post Octavo, price 7s. 6d.

---o---

HISTORY OF FRANCE,

FROM THE EARLIEST PERIOD TO THE YEAR 1848.

By the Rev. James White,
Author of the "Eighteen Christian Centuries."

Post Octavo, price 9s.

"Mr White's 'History of France,' in a single volume of some 600 pages, contains every leading incident worth the telling, and abounds in word-painting, whereof a paragraph has often as much active life in it as one of those inch-square etchings of the great Callot, in which may be clearly seen whole armies contending in bloody arbitrament, and as many incidents of battle as may be gazed at in the miles of canvass in the military picture-galleries at Versailles."—*Athenæum.*

www.ingramcontent.com/pod-product-compliance
Lightning Source LLC
Chambersburg PA
CBHW032009300426
44117CB00008B/952